AF454342

Three Fire Mountains

Katie Simmons

Three Fire Mountains

Stories of Wildfire and Recovery in California

 Springer

Katie Simmons
Stanford Impact Labs
Stanford University
Stanford, CA, USA

ISBN 978-3-032-17342-3   ISBN 978-3-032-17343-0 (eBook)
https://doi.org/10.1007/978-3-032-17343-0

This Springer imprint is published by the registered company Springer Nature Switzerland AG
The registered company address is: Gewerbestrasse 11, 6330 Cham, Switzerland

If disposing of this product, please recycle the paper.

*To my parents and all fire survivors,
especially the Paradise ladies.*

Foreword

My co-worker was a 911 dispatcher for twenty years and calls the Camp Fire the best and worst day of her career.[1] Heroism and horror. I consider leaving twenty-five-years in non-profits to work in public sector disaster recovery the best and worst decision of mine.

I set out to write a clean, concise paper for Stanford University, offering tips and tools for working in recovery. I envisioned using my relatively new knowledge and experience to light a path through the wilderness of wildfire, making the journey simpler and less painful for others. I used to say recovering from the Camp Fire was the hardest, most expensive process imaginable because nobody had done it before. Now that I've lived through additional fires with different complications, and witnessed wildfires in other areas bringing their own unimaginable challenges, I see how futile it is to try to make recovery less impossible.

I find hardship less hard when I say out loud, "this is hard." I admire people who do hard things when they admit how hard they're working or how hard the work is on them. I no longer believe the purpose of this project is to make recovery less hard, I believe it is meant to show the suffering so that when others suffer through this they're not suffering alone. As tempting as it is to hold up our successes for recently destroyed communities to see, it is far more honest to pull back the curtain to expose our own pain and suffering as we cobble together long-term recovery.

This book contains descriptions of animal pain and death, my suicidal thoughts, and my dad succumbing to alcoholism in pursuit of recovery. I thought being a Policy Fellow meant I'd keep my topics safe and contained—data-driven, as they say—but wildfire has a way of inserting its wildness into just about everything, and I'd be faking my entire disaster recovery experience if I pretended otherwise.

I feel like I've given up everything to this profession but have gained the whole world. I can no longer access my diplomatic, yielding, people-pleasing self. Recovery rage burned down my social buffers and left my personal relationships smoldering. I missed a dear friend's birthday recently and didn't realize it for a month. I am not tied to people in the same way I used to be, outsourcing my thoughts and feelings to

---

[1] Fellowship interview, Angie Mannel, Butte County Emergency Manager, 2025.

relative strangers on social media. I have retreated and find myself alone on Trauma Island, surrounded by the toughest, harshest, most sweeping feelings I've ever had. I'm here by myself, in my own body, enmeshed fully in my own thoughts, and I've never felt more lost nor found.

Wildfire wrecked me. Or did it?

I was browsing with my daughters on the second floor of a local antique store when a woman's voice floated up. "My house didn't burn down," she said, matching the sorrowful tone of the conversation. I realized then we're in an era when houses are burning from dozens of fires, not just one identifiable tragedy. In the summer of 2024, we held information sessions for survivors of four small fires had that recently occurred. The bigger fires have their own meetings. In the fall of 2024, just off of the Park Fire, we hosted a delegation of housing representatives from the State who, after reviewing the list of burn scars to visit, decided on the most recent to witness debris and tree removal rather than reconstruction.

Fire. Is. Everywhere.

And yet.

Having put myself through the paces of the fellowship, I know what keeps me motivated. I know my definition of recovery. I know why I keep showing up. I don't need to interview folks from across the state and read books and articles to find the answer—I need only the bravery to look inside of myself and be honest about what I find there, and then share it with others.

After the LA firestorm in January of 2025, a member of the Butte County Board of Supervisors and I drove to Sacramento where we met with Assembly member Harabedian's staff in a tiny, sunlit office. We sat looking at their stricken faces, realizing they didn't know the first questions to ask. At the outset, it seems logical to start with week one, then proceed into the sequential steps people need to take to put their lives back together. Tell us where to go from fire, they seemed to be asking, as if we could unveil the linear minutia of a perfect recovery process. As if we had it square in our own minds.

To avoid revealing too much madness, we spent that precious hour talking in big, definable buckets: insurance, debris removal, local ordinances. We offered insights into the big picture and tried earnestly to normalize their sense of overwhelm. I realized quickly they only needed two things from us: (1) proof of survival, and (2) our contact information.

When facing complex tasks, it makes sense to start at the beginning and tease them apart. With fire, the beginning is mottled with confusion, grief, trauma, and blurry thinking. The fits and starts of early recovery can lead to finger-pointing as the upside down, backwards process begins to reveal itself. I told a philanthropic professional in LA who'd been tapped to lead charitable giving for wildfire recovery that everyone wants to hear it's A, B, C, D, E, F, G…, when it's actually N, -2, purple, 16, apple, B, winter, 88 divided by 153, X on Monday, and the reverse on Friday.

Federal and State government agencies publish documents on the topic of recovery; long, complex, technical instructions written in acronyms. If you can make it to page six of the Federal Emergency Management Agency (FEMA) National Disaster Recovery Framework you'll read, "The FCO serves as the counterpart to lead state, territory, or Tribal Nation response official and has primary responsibility for coordinating federal disaster response and recovery support to the whole community in accordance with the NRF, NDRF, and Federal Interagency Operational Plans (FIOPs)."[2] Flip back to page three and you'll see the NRF is defined as the National Response Framework. The NDRF is defined on page one as the National Disaster Recovery Framework. The paragraph above the quote describes the Federal Coordinating Officer (FCO), the Federal Disaster Recovery Coordinator (FDRC), and the Federal Disaster Recovery Officer (FDRO). Turn back to page one to remember SLTT means "state, local, Tribal Nation, and territorial governments," and then try to remember the differences between the FCO, FDRC, and FDRO in the context of the NRF, NDRF, and FIOPs.

I don't believe disaster recovery can be taught by a government manual, but I know it can be learned. When onboarding staff, I've found the best way to humanize and explain recovery is to stand before the whiteboard in our conference room and draw a mountain. The mountain is what recovery feels like as various government agencies go through the motions of appropriating a congressional allocation for a federally declared disaster. Eventually, about halfway up the mountain—three years in—those motions trigger the very first steps toward funding local recovery. The whiteboard cuts to the chase of a decade of Action Plans and amendments, notices of intent, applications, standard agreements, and close-outs in a "you are here" kind of way. Regulatory reading is required before administering federal funds for disaster recovery, but it's not a map. We can't put communities back together with jargon, nor use it to light our way.

Recovery is complicated enough without layering in regulatory compliance, and it is shocking what the manuals don't say about the realities of recovery. Peppered throughout FEMA's National Disaster Recovery Framework, Third Edition—Amended, are success stories resulting from the framework model in action. As a recovery professional, I don't need success stories, I need to know how destroyed communities facing impossible trade-offs keep their heads up.

I'd like to say out loud what I've thought so many times and leave it at that: recovery ruins people and places as much as fire does. But that's a deep oversimplification that dishonors the work so many people, including myself, are putting into it. Recovery does not restore communities to their pre-fire states which, even if it did, would not erase fire from the lives of the people who survived it. Nor does it bring back the people who did not. When you strip away the politics and the acronyms, recovery is an instinctive human reaction to enormous tragedy; a constant surge of somewhat blind, heartbroken activity toward homeostasis. I recently described it as an ultramarathon for sprinters and the whole crowd of recovery professionals and volunteers nodded.

The truth is, at its very core, recovery is a feeling. It is the motivation behind every act of kindness following despair. It is what compels us to drive to lawmaker's offices in Sacramento to talk staff through their grief under the guise of assistance. It

---

[2] FEMA National Disaster Recovery Framework, Third Edition—Amended, December 10, 2024, https://www.fema.gov/sites/default/files/documents/fema_national-disaster-recovery-framework-third-edition_05062025_0.pdf.

is what makes us sit with survivors from other communities and helplessly listen to them weep. As much as I want to jump to the end of this book and wipe the pain of experience from my hopeful conclusion, I don't believe you can understand recovery without knowing what comes first.

Fire.

Chico, CA, USA   Katie Simmons

Acknowledgements

This story would still be trapped deep in my mind without Stanford Impact Labs and their first Policy Fellowship for local government professionals. The fellowship and my mentors pulled the pieces of this tale out ever so gently, guiding me along the research path, and allowing me to stray into anecdotes, opinion, and occasional tears. Thank you to Hana Passen for creating the fellowship and somehow finding my email address, Michelle Skoor for making it ok to stumble *and* enjoy the sunset, Ashima Tshering who took my "wheel of cheese" and turned it into the recovery triage tool, and Amanda Grayor who said, "your story matters," at just the right time.

Participating in the fellowship wouldn't have been possible without my boss, Meegan Jessee, and our Chief Administrative Officer, Andy Pickett, who let me travel to Stanford University literally and figuratively to explore this topic outside of what I do day in and day out for Butte County. The fellowship paid for my staff time and handful of days away from the office. Thank you.

Debbie Johnson, the guardian angel of my recovery heart and my one true reader, knows she made this happen by editing, commenting, and texting at just the right moment in my writing journey. As I've told Debbie many times, if she's the only reader, and this book helps heal us, then it'll have been worth it. Thank you, Debbie, for delicately helping survivors navigate the retraumatizing and enraging process of recovery. Only you can do what you do.

The staff in my office who push the thankless plow of recovery in the barren field of disaster are unseen heroes in my eyes. Every single day we run up against the craziest, toughest, most ridiculous barriers in an effort to do the impossible. Through tiny daily miracles, we somehow manage to invest federal dollars into local communities. We do this together and I'm endlessly grateful to everyone who has shown up for this.

To my friend, Caryn, thank you for recognizing my secondary trauma and giving it a name. You spent many long, tearful hours on the phone with me as I navigated my way toward starting to understand my complex heartbreak. You allowed the cow story to come out, in single, weepy words until it was an art project that still gives me joy. You got in the trenches with me, held my hand, and helped me start crawling out of the darkness.

Dave Daley, a cattle rancher in Butte County, will likely never know how sharing his search and rescue story on the Butte County Farm City Tour in 2022 would change my life. His anguish made space for mine. It's taken me years, but I now know the precise color, shape, and size of my pain because he was brave enough to share his.

The Paradise Ladies—Melissa, Judy, Pam, and Kelley—forever my wise and generous friends. They each lost their homes and with their own hands and hearts helped Paradise take its first steps forward. Colette, Marc, Kevin, Kate, Crystal, Lauren, Susan, Dina, Aisha, and Vickie, too many others to name—I am so lucky to have spent that one tragically beautiful year with you at the Town of Paradise. You will always be the strongest people I know, hands down.

To the many community members who reached out with honest feedback and edits along the way—thank you. This fellowship allowed them to see me in new ways, and to see the many ways in which we suffer together. I'm so sorry this is our bond, but this battle is tolerable because of you. And to the consultants, vendors, reporters, and scholars who are interested in what we do—thank you for your calls and questions and the opportunity you give me to better articulate how to treat wildfire survivors. We don't always have the easiest conversations, but the result is always gentler, more equitable treatment of people at their most vulnerable.

Finally, my family: my parents who were thrust into the heaviest, adrenalized days post-disaster and who managed to care for my kids in the process. My sister who let us stay on her couch that first devastating night when the last we saw of our community were great mountains of uncontrollable flames. My aunt who let us find clean blue skies at her cabin in Tahoe during those first weeks, giving the kids a sense of normalcy while ashes and smoke lingered for months in our air and on every surface. My cousin who checks in on my writing process, and my uncle who let me stay in Santa Rosa for a week of pure editing. Ah, bliss. Our cats—Mim and Lulu—are the best late-night writing partners, and Sadie, our rescue Great Pyrenees, continues to fill my heart and ease my big animal pain.

To my husband and daughters, to my community, we are warriors.

Introduction

This book was prepared over the course of a Policy Fellowship for Stanford Impact Labs between November 2023 and October of 2025.[3] Stories are personal accounts of living and working in wildfire and recovery in Butte County, California, located 70 miles north of Sacramento in the foothills of the Cascade and Sierra Nevada Mountain Ranges.

Interview material was collected from willing individuals and agencies who shared their experiences and perspectives understanding they might appear in fellowship outcomes. Data and graphics are from publicly available information. Anecdotes about individuals, including case studies, are presented without personally identifying information. Instructional material comes from the Butte-Glenn Community College Training Place and their sources, except as otherwise referenced.

---

[3] Seven Staff Members in Local Government Take Part in Brand New Stanford Fellowship, Stanford Impact Labs Blog, November 27, 2023, https://impact.stanford.edu/article/seven-staff-members-local-governments-across-california-take-part-brand-new-stanford.

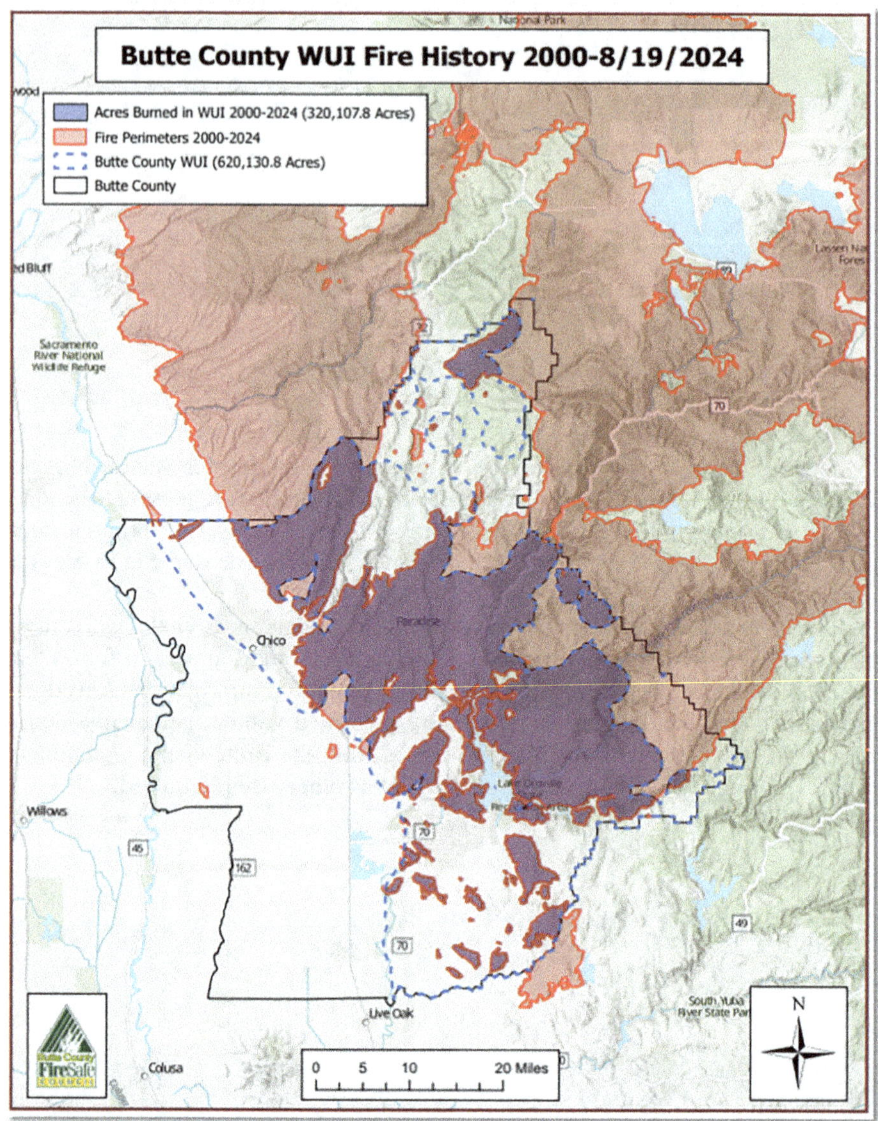

Fig. 1 Map of fire activity in Butte County from 2000 to 2024, prepared by the Butte County Fire Safe Council

The intent of this book is to examine a variety of experiences to explain the hardships of wildfire and recovery, focusing on funding, programs, and processes meant to achieve community viability after disaster. All opinions expressed are my own and do not express the opinions of Stanford University, Butte County, or the Town of Paradise.

Recent Butte County Disasters

- 2024—Thompson Fire, Park Fire
- 2021—Dixie Fire
- 2020—North Complex Fire, COVID-19 Pandemic
- 2018—Camp Fire
- 2017—Oroville Spillway Incident, Wall Fire, Ponderosa Fire, LaPorte Fire.

Contents

Common Disaster and Recovery Acronyms

CAL FIRE	California Department of Forestry and Fire Protection
Cal OES	California Office of Emergency Services
CalRecycle	California Department of Resources Recycling and Recovery
CDAA	California Disaster Assistance Act
CDBG-DR	Community Development Block Grant Disaster Recovery
DINS	Damage Inspection Report
DROC	Disaster Recovery Operations Center
EOC	Emergency Operations Center
FEMA	Federal Emergency Management Agency
FMAG	Fire Management Assistance Grants
HCD	California Department of Housing and Community Development
HMGP	Hazard Mitigation Grant Program
HUD	United States Department of Housing and Urban Development
IA	Individual Assistance
ICP	Incident Command Post
ICS	Incident Command System
IMT	Incident Management Team
NIMS	National Incident Management System
OOR	Owner Occupied Reconstruction
PA	Public Assistance
PSPS	Public Safety Power Shutoff
SEMS	Standardized Emergency Management System

Chapter 1
The Fires

Recovery begins before fire even starts.

It is shaped by who we are, how we feel, and what we think. Recovery is made up of the stuff that gets into our bones when we're children, and defines how safe we feel and what choices we make. For me, recovery began in kindergarten.

Petaluma, California, is north of San Francisco near the Pacific coast. The fog lays a briny layer over the hills which, when they warm up, steam with manure. In the mid-seventies, my parents brought me home to a single-wide trailer with wood paneling and shag carpets. I remember thick, crocheted wall hangings and furniture in shades of green, yellow, and orange. Memories smell like boat fuel, sawdust, and alcohol.

Shortly after my parents moved out of the trailer and into the house my Papa co-signed for them on Kresky Way—with shellacked pine floors and stained-glass window—my dad moved out. They shuffled a few more times in the negotiation of divorce, my sister and I going back and forth between them, before my dad moved back to Kresky Way where he died forty years later. The Tubbs Fire delayed the close of escrow when I sold his house in 2017.[1]

Midway through kindergarten, before we moved out of town with my mom, I found a magazine in a friend's bathroom. She lived across McDowell on a corner in Petaluma and we had sisters the same age. I remember wondering why her parents fought louder than mine but stayed together.

Staring back at me from the magazine cover were children with no performative expression. They were emaciated in very little clothing—maybe not children at all—and I felt shame studying their bones, eyes, and teeth. I flipped through the pages thinking there must be some mistake, these photos couldn't possibly be real. Seeing suffering on a scale I didn't know existed bewildered me, and I felt the first of many tugs to act on behalf of somebody else's pain.

DO SOMETHING, the photo pleaded. Nothing, my young body replied.

---

[1] Tubbs Fire (LNU Complex) CAL FIRE Incident Report, 2017, last updated 10/25/2019, https://www.fire.ca.gov/incidents/2017/10/8/tubbs-fire-central-lnu-complex.

Four decades later, a year after the Tubbs Fire shook Santa Rosa, the Camp Fire turned that tug into a life-changing decision. That magazine cover may be the whole reason for this story.

Camp Fire, November 8–25, 2018

There is no preparation for witnessing tragedy unfold, for the shock and horror of taking in the unimaginable. I had watched my share of big-screen movies by the time I saw the Camp Fire, but nothing prepared me for an entire mountain in flames nor the fear the fire may reach my children before I did.

Humboldt Fire, June 11–21, 2008

A decade before, I'd driven through fire with my three-year-old in the back seat. It was 2008 and the Humboldt Fire was racing toward Paradise in a billowy blue-black cloud. After picking up Adela from preschool I had her wait in the car while I grabbed a few things from our rental. I forgot "Naked Baby" which is how Adela ended up with two baby dolls when I bought "Bee Baby" in Target a few days later.

That afternoon a thick line of cars was gridlocking its way up the ridge. Because I could see the smoke downhill, I wondered why people were moving in the same direction as the fire, heading straight uphill where the roads got narrower and eventually turned to gravel. I didn't think at the time that remaining on the perimeter of fire and moving faster than it could be the best and only strategy, nor that law enforcement might be directing people that way. I had no concept of how or why a community moves during fire, I only thought about getting myself and my child into the valley where we'd have options. At this point, I was three months into life in the Sierra Nevada foothills thinking surely the best route out was down.

As I moved toward my decision point at the intersection of Pearson and Clark Road, a sedan slipped in front of us with a goat wedged in back. I pointed this out to Adela in my sing-song mom voice, then rolled down my window and shouted across to the cars, "is the road still open down there? Can we get out?!" It was the kind of thing you hope never to shout within earshot of a child. Can we get out? Will our car be engulfed in flames? Will we die? "I don't know," someone shouted from a stopped car pointing uphill, "if it's open now it might be closing soon!" I stepped on the gas and drove down the hill, the only car on a lone mission.

Driving into flames causes an instinctive push–pull. The body says no, the mind says yes, then the body says yes and the mind says no, then all systems are overtaken by indecision and a robotic human energy takes over. At a certain point there is no direction but forward. As the saying goes, the only way out is through.

This moment plays back slow-motion in my memory. The fire was a delicate, bright filigree on the edge of the pavement like the rim of the setting sun just before it dips below the horizon. Above, flames rose into a hinging jawbone madly chewing up the hillside. I drove in regret and determination, weeping all the way down.

Adela was not a car sleeper but knew to close her eyes when I ran out of stories to explain this away. With the trees on fire, there was nothing more to say about this place I wanted us both to love.

Down the hill, firefighters were too busy to worry about my car sneaking through, though I vaguely recall arms waving. They were protecting the community college at the base of the hillside in a grassy valley. On the outskirts of the fire, the grass perked up and the oaks stood ready. I have since learned that fire pushes forward with its own wind, warning everything in its path. It was June in Northern California, after all, peak golden hill season with fuels on every surface.

I tapped my fingers on the steering wheel and counted in my head, a life-long habit of metronomic soothing. Beat for every street sign we passed. Beat for every intersection. Beat for each fence post. Beat beat beat, tap tap tap as the white dashes sped past. Adela slept; her eyelids shut tight as if pulled from the inside.

And then…the smoke was behind us, still visible roaring up the mountain, blowing away from us. I made a call. I was heading toward my mom and stepdad's house in Chico where they'd moved from the Bay Area in 2007. When I arrived, no one seemed to understand what we'd just been through and I was probably too shocked to explain. Years later, as fire became more frequent and severe, the experience would sink into my gut like an immovable rock, but that afternoon it was nothing more than a shocking way to pass the day. To shift gears for Adela, we headed to the farmer's market.

The black smoke plume was visible from downtown Chico that evening where we walked past vendors and food stands, a strange sunny oasis outside the fiery shade. In all, 254 residential, commercial, and other structures were destroyed, 10 firefighters and civilians were injured.[2] My brand-new town was burning but not as badly as it would 10 years later. Not nearly as badly as it would, the Humboldt Fire scorching a powerful path for its successor.

Thankfully, toddlers have a way of lightening the mood. When Adela said "ash" it sounded like "ass" and she was not a quiet child. At home, in the car, at preschool, at my office, in the grocery store, in front of everyone for the duration of the fire:

It smells like ass!

There's ass everywhere!

I have ass in my hair!

In all fires since, when my family talks about "ash" it becomes "ass" because it's the only way to cope. Fire season smells like ass.

1.1 Midnight Morning Sun

On November 8, 2018, I was four days into a new job in Sacramento 70 miles from home. I'd left a job I loved after 7 years in Chico and was commuting an hour and a half each way. My husband texted that there was a fire, a smoke cloud was growing in the east toward Paradise and taking over the sky. Adela had texted him while

---

[2] CAL FIRE Incident Report Humboldt Fire, last updated 6/21/2008, https://www.fire.ca.gov/inc idents/2008/6/11/humboldt-fire.

walking to junior high that morning wondering if everything was ok. Her school sat adjacent to the grasslands that back up to the hills where the smoke was black not brown, the plume wide not thin. He assured her the fire would be out soon, nothing to worry about, but by the time he texted me he wasn't so sure. Just down the street from Adela's junior high, Heidi was in 1st grade.

My mom started texting photos of charred debris she held in her hand like a crow's wing. She and my stepdad now lived in Butte Creek Canyon, a steeply-walled valley below Paradise on a stretch of peninsula flanked by the creek. In her texts she marveled at these handfuls of ashen rain, giant flakes of life in Paradise falling from the sky, a horrifying alert to anyone upwind.

"MOM," I texted back, "GET OUT." By now I could hardly concentrate on work. She would eventually evacuate by order of the Sheriff along a bike path that connected her canyon to the main road heading into Chico from Paradise, the road that, by this point, was operating in contraflow for the mass evacuation of the ridge.

I was hesitant to leave early from work because I was so new to the company. Hell had arrived in the North State but I was on the clock and no one else's phone was blowing up like mine. When I finally left the office, it was with direct instructions to the family: don't wait for me if you need to get out. JUST GO.

The most direct routes heading north from Sacramento were closed. Highway 99 through Chico and Highway 70 through Oroville were being used for fire management. I cut west to I-5 and kept north, wondering if I would eventually be able to head east toward the hills, toward the fire. Seeing no road closures I chose the backroads into Chico where, sure enough, thousands of headlights raced past me on the two-lane road escaping dodge. Once again, I was driving against traffic toward fire.

Dozens of miles out I could see the blaze and my texts turned to panicked calls. "Honey!" I shouted, "it looks like the fire is in Chico! Why are you still at home?! Take the kids and get out!" Our home is on the southeast side of Chico where the fire would eventually come within a mile. It was a call I'd make several more times before arriving home. Adela started calling me from her phone about evacuation packing because she'd been through this before. At first, I told her to pack nothing for me then at her insistence I listed off quick things she could grab so we could turn around and get out as soon as I got home: my robe, my earplugs, my shoes, my running shorts. I was preparing for sleep and exercise; I could think of nothing else.

When I arrived home, Adela was ready with her photo albums and sentimental treasures packed neatly in plastic bins. When she moved out for college several years later her treasures were still packed from our Camp Fire evacuation. Heidi's "friends" overflowed from several expandable duffle bags and she was ready to go. We did several checks for Big Kitty to make sure she didn't get left behind like Naked Baby, because there is no replacing a 35-year-old stuffed animal Heidi inherited from Rodney's mom.

In a decision he'd later regret, my husband hung back. We made a rushed and shouty plan that if the fire got too close, he'd leave. I begged him not to hose the roof, not to sprinkler the lawn, not to try to fight the fire. You don't have the skills, I pleaded, don't risk it.

We waved a long time before the girls and I entered the long line of traffic bumper-to-bumpering back to I-5 on the road I'd just come in on. All I remember from snaking through the orchards was not knowing if I'd turn north or south at the interstate. I decided to go the way traffic was lightest to get as far away from the fire as fast as I could. Eventually we made it to my sister's house in Sacramento where my kids and parents would stay on and off for the next several weeks. We looked like ghosts after the apocalypse in photos from that night, the girls spooning on the couch, holding each other in sleep.

It would be weeks before we knew that my parents' house survived but their land and outbuildings did not, and months before we knew the extent of the destruction despite the fact that most of it occurred within the first few hours. The fire moved at record-breaking speed from Pulga toward Paradise, pushed by unyielding winds across the treetops, dense vegetative fuels, low fuel moisture, and sloping terrain.[3]

PG&E had seen danger in the forecast and had scheduled a power outage that day but called it off in the early morning before ignition. It is beyond me to consider what would have happened to the people on the ridge if the fire had started any earlier.

Wildfire can smolder inside of tree trunks for weeks, fueled by what it finds underground. Daily containment percentages are wicked reminders of how long fire can linger. The fatality count climbed as folks moved from the missing column to the deceased column or, halleluiah, to the found alive column. Pets were transported to shelters, many of them seriously burned, then to massive facilities where some lived out their final days unclaimed. Large barn animals and wildlife found dead were transported to the landfill where, as I later learned as part of my disaster recovery duties, a permit had been obtained to receive large animal carcasses.

Oh, the destruction, oh the devastation. It still plays in my mind like a melancholy song. There would be a criminal trial, 84 counts of involuntary manslaughter, heaps and gobs of promised money, complaints filed with the California Public Utilities Commission, lawsuits threatened against the most victimized, and deep mental, emotional, and physical scarring and displacement that lasts to this day.[4]

On our street in Chico after the fire, a neighbor paved her front lawn where an RV parked for several years with patio furniture just outside the swinging side door. During this past Christmas season a homemade sign illuminated by walkway lights said, "Remembering our home in Paradise," on a front lawn around the corner. Our next-door neighbors lost their home in Paradise as did the woman four doors down.

In all, 85 civilians died in the Camp Fire, most of them in Paradise. The fire destroyed 18,804 structures, damaged another 754, and injured 3 civilians.[5] It would be these numbers, compounded by subsequent fires, including the state's most destructive fire in 2020 in Butte County, that would consume my professional life until now.

---

[3] Maranghides et al. (2021).

[4] Lagos and Jamali (2020).

[5] CAL FIRE Incident Report, Camp Fire, last updated 9/26/2024, https://www.fire.ca.gov/incide nts/2018/11/8/camp-fire/.

Housing is the focal point of all efforts, all energy in recovery. For populated areas, recovery must begin with housing even if the houses do not come first. Without housing there is no economic development, no provision of services, no community life, no intangible sense of place, no return to home.

Before housing, though…cows.

1.2 Mountain Herd

North Complex Fire, August 17–December 3, 2020

Dave Daley is a 6th generation cattle rancher in Butte County.[6] Before the heat of summer, he drives his herd east into the mountains where the grass is still green. His cattle roam the wooded mountains until they are driven back down into the valley to graze on grasslands in the winter.

The North Complex Fire broke out in late summer of 2020 when COVID was in full swing and the community was still reeling from the Camp Fire. I was Disaster Recovery Director for the Town of Paradise at the time and my duties ranged from hazard tree removal to wrapping up the After Action Report. One particular morning we awoke to a blackened sky that didn't lighten as the sun rose; rather, the blackness turned to orange, a telltale sign of thick ash in the air from a nearby fire. The trauma was palpable. I called my parents from my drive into Paradise that morning and said, "don't wait this time, get out."

When I arrived at work, I was asked to secure the building with a co-worker, so we circled the perimeter of Town Hall to see what flammable materials we could pull away from the structure to avoid ignition. I felt ridiculous stepping over knee-high weeds in my shiny work flats, sweater, and pants while listening to my colleague's instructions. I wasn't dressed for fire and neither was the building.

Back inside we gathered in the Town Manager's office where the former Fire Chief, Jim Broshears, who still worked as a volunteer emergency coordinator papered the table with maps and leaned heavily into them with his palms flat. As he called out the names of ridges and peaks I'd never heard of, he traced his finger along the topographic lines to show where the fire was heading. He explained how the fire would accelerate in the steeper slopes of the canyon and where the wind might divert it from the ridge tops. From the windows we could see the glow of the flames a few ridges over, fire's false sunrise.

I later learned the necessity of maps during fire. In my current emergency response duties, I shuttle maps between first responders and local government employees coordinating mass care and shelter, road closures, and services for the affected population. I remember how Chief Broshears needed those maps to articulate fire behavior and probability so the Town could prepare. That morning, for a newcomer to fire like me, the maps made it possible to understand the fire.

---

[6] Morgan (2024).

PG&E implemented a planned PSPS event in the Town that morning—a Public Safety Power Shutoff of their utility lines to avoid the danger of ignition caused by high winds—which meant that except for the buildings and street signals operating on generators, the town was dark. I went in and out of Town Hall in restless agitation and watched a persistent line of tail lights head downhill toward Chico, red lights glowing in the ashen dawn fading at the bend. No one was waiting to be told to get out this time, to be caught off guard by fire again.

Steadily the town emptied out: fifth wheels, horse trailers, RVs. The intersections without back-up generators were dark and the public's exit was careful but insistent. We eventually called an Evacuation Warning for the whole town, heeding the signal residents were clearly giving. No one was taking chances because there were no chances left to take, the worst had already happened. I remember standing on the side of Skyway outside of Town Hall under the darkness of mid-morning fire, Camp Fire destruction all around us, watching the cars leave town, my eyelashes filling with ashes from a new fire, feeling like my insides were burning down.

It was during that fire that Dave Daley's cows died while grazing in the mountains.[7] The North Complex Fire started in August and he hadn't yet driven them down for the winter. The cows were familiar with the area and allowed to roam free.

Dave took photos from his rescue mission a few weeks later where he found piles of cows painted thickly with ashes and soot—strong black cows turned gray and thin.[8] On that mission he found a pregnant cow standing in a puddle attempting to sooth her charred legs. Because she was injured beyond saving but her calves were still alive, Dave did what ranchers do and relieved her of her misery. He tried to save the calves but they did not survive the ordeal and he pulled them out dead, one by one.

I heard Dave tell this story a few years after the North Complex Fire. As he described the scene, I felt my brain rewiring to take in more unbearable suffering, images I'd never imagined but could not unsee in my mind's eye. This is the reality of living in the aftermath of fire, every encounter carries the potential to break your heart.

Dave made a few recoveries on that mission, a handful of rescues, but as any rancher will tell you, cows that have been through fire may never fully recover. The experience is too much for many to survive unscathed and eventually thrive. As I've read about the impact of wildfire on cattle ranches in other parts of the state, I know that many cows who've been through fire are weakened by stress and die prematurely.

Back at Town Hall, standing under the falling ashes that morning, I couldn't have known what was going on in the mountainous flames a few ridges over. The North Complex Fire never reached the town but it took 16 lives, more than 300,000 acres, and well over 1,000 homes in quaint remote communities of Butte County.[9]

---

[7] Dave Daley shared his story at the 2022 Butte County Farm Bureau Farm City Tour 2022.

[8] Daley, D., I Cry for the Mountains and the Legacy Lost, The Bear Fire, California Cattlemen's Association, https://calcattlemen.org/2020/09/23/legacy/.

[9] Mapping of the North Complex Fire, UCMapping 4, July 10, 2022, https://storymaps.arcgis.com/stories/e905b0aa9d224c95ac8fa7b579cbb66e.

Jim Houtman, a retired firefighter who now works for the Butte County Fire Safe Council told me so many animals were lost in that fire that CAL FIRE began keeping count of wildlife fatalities, a practice never done before.[10]

From Town Hall that day I only knew we had an emergency to run, my first real Emergency Operations Center (EOC) experience, one I would repeat my first week on the job at the County where I moved to release myself from the pressure cooker of recovery work. My first day at the County was July 12, 2021.

Dixie Fire, July 13–October 25, 2021

The Dixie Fire ignited in Butte County and gripped the North State for several months, eventually scarring almost a million acres in several counties, widely reported by CAL FIRE personnel as the largest single-ignition fire in state history.[11] During those months the Dixie plume was visible from our front yard, stealing our summer and peace of mind.

My then 8-year-old perfected her rain spell during the Dixie, a fire so big it created its own weather. She found the instructions in her fairy book embossed in gold, put three blades of grass in a glass of water, stirred to the left three times, stirred to the right three times, then poured the grassy water outside on a tree trunk. The spell said this would bring rain and it did a few times. After a while we didn't even need the book to remember the spell. I remember it still.

During the Dixie EOC I served as Liaison. This means I went back and forth between the Incident Command Post (ICP) in Chico where the firefighters and responders were stationed as the Incident Management Team (IMT), and the County EOC in Oroville where County staff were working on mass care and shelter plans, communications, public works response, and other logistics. After each IMT briefing, I brought incident maps and Incident Action Plans to the EOC. As the primary contact for the IMT Liaison, I coordinated their report-outs on our briefings so County staff could hear directly from CAL FIRE who was jointly operating with the United States Forest Service for the Dixie Fire. Though the fire started in Butte County it moved well beyond it, eventually requiring a second Incident Command Post in another county to account for the closed roads in between.

On Saturdays and Sundays, County staff assigned to the Liaison role rotate going to the ICP for briefings. When it was my turn, I took my daughters with me. We brought fire maps home that my youngest used for firefighting make-believe, moving her plastic dinosaurs around the mountains, surrounding the fire perimeter, putting out the fire by splashing in the lakes and streams. She made sure firefighting aircraft was parked on the scene at all times.

Currently, my job at the County is to supervise emergency management and the administration of disaster recovery funding from federally declared disasters. Because the role was focused on economic and community development before

---

[10] Information provided by Jim Houtman, Butte County Fire Safe Council and former fireman for the City of Chico, CA, 2023.

[11] CAL FIRE Incident Report, Dixie Fire, last updated 10/14/2022, https://www.fire.ca.gov/incide nts/2021/7/13/dixie-fire/.

it morphed to include disaster recovery, this includes housing and public service grants. For many years, the division was responsible for traditional Community Development Block Grants (CDBG) and, most recently, COVID grants. When I arrived at the County our grant-funded programs ranged from long-term loans for housing rehabilitation, food bank distribution, housing navigation services, mobile hygiene services, and business assistance grants. With American Rescue Plan Act (ARPA) funds allocated to Butte County, we oversaw Board-selected projects like recreational trail construction, broadband planning, wayfinding signage, for-profit business grants, and special event grants to offset losses due to event cancellations caused by the pandemic. In the simplest terms, grant-related work for economic and community development requires securing federal and State funds to subcontract out to community organizations to deliver Board-directed services to the public.

Toward the end of 2022 my boss announced a transition at work. I'd be assuming supervision of our long-term disaster recovery grants as I did at the Town of Paradise. These grants are considered last-in unmet need funding meant to restore public infrastructure, build multi-family housing, acquire and operate facilities and County services, as well as mitigation planning following disaster. Due to the severity of the Camp Fire, these recovery grants are allocated at over $245M to the County jurisdiction over a span of about a dozen years. Projects funded by these grants are prioritized by the Board based upon how the funds can be used, when the County must use them, and where the community needs them most.

During the fall of 2024, I began supervising the County's emergency management staff administering response and recovery programs, including cost recovery and insurance collections.

References

Lagos M, Jamali L (2020) PG&E pleads guilty to involuntary manslaughter in deadly camp fire. KQED—The California Report. https://www.kqed.org/news/11808166/pge-pleads-guilty-to-involuntary-manslaughter-in-deadly-camp-fire

Maranghides A, Link E, Brown C, Mell W, Hawks S, Wilson M, Brewer W, Vihnanek R, Walton W (2021) A case study of the camp fire—fire professional timeline. Technical Note (NIST TN). National Institute of Standards and Technology, Gaithersburg, MD [online]. https://doi.org/10.6028/NIST.TN.2135. Accessed 9 Oct 2025

Morgan T (2024) One year after wildfires decimated Rancher's herd and legacy, devastation fuels change. AG Web Farm Journal. Last updated 30 Oct 2024. https://www.agweb.com/news/livestock/beef/one-year-after-wildfires-decimated-california-ranchers-herd-and-legacy-devastation-fuels-change

Chapter 2
Early Recovery

When I first started disaster recovery work, I noticed something surprising: early recovery feels hurtful to the community rather than helpful, or at least hurtful while it's helping. Recovery, as I've come to find out, is the necessary wounding for healing to take place. It is the final hand of destruction before rebuilding can begin.

I oversaw the hazardous tree removal program in the Town of Paradise after the Camp Fire. For me, the title Disaster Recovery Director initially conjured up words like restoring, rebuilding, re-establishing, re-creating, and re-invigorating to describe what I thought the work would be. But that was not the case. Early recovery is removing everything fire has left behind, then taking what disintegrates during the aftermath…if resources are available…until literally nothing is left.

Household hazardous materials melting into the ground during the Camp Fire required the removal of several layers of earth along with debris in order to remediate residential and commercial lots.[1] State-led debris removal crews contracted by CalRecycle left gaping holes where homes had been, and chain link fencing around what wasn't eligible for removal like swimming pools.[2]

Homes can't be rebuilt in holes so lots have to be filled and leveled before reconstruction can begin. To this day, some Camp Fire impacted property owners are on a waiting list for PG&E to deliver free or reduced-cost soil to level their lots.[3] Soil that's removed after fire can sometimes be cleaned and re-used, but that's neither an easy nor quick process. Remnants must be taken out, the soil finely combed then cleaned and sanitized to the point of hazard elimination. I picture this involving a giant sifter like the one my grandma used to add flour to her pie dough, a puff of soil with each shake.

---

[1] Observed processes and decisions while sitting on the Incident Management Team for Camp Fire Hazardous Tree Removal Operations, Cal OES, 2020–2021.

[2] Eligibility requirements announced by Cal OES and CalRecycle in public information provided during Camp Fire Debris Removal and Hazardous Tree Removal Operations.

[3] Community discussions with PG&E personnel during Upper Ridge Community Council Meetings, Magalia, CA (2024).

K. Simmons, *Three Fire Mountains*, https://doi.org/10.1007/978-3-032-17343-0_2

Removing trees through a government-funded process is also not as simple as it sounds. In hazardous tree removal programs, dead and dying trees must be found eligible to be removed. Of all the recovery programs I've worked on, tree removal is the most complex and, in my experience, the most emotionally taxing for survivors living in wooded areas.

Eligibility is enforced by State agencies and involves a somewhat complicated process of employing arborists to determine a tree's diameter and height in relation to a publicly-owned or qualified private roadway. Under the Camp Fire program, removal wasn't about cutting down trees that threatened people or property, it was about removing trees of a certain size that threatened the public right of way. And removal wasn't complete—stumps were left to a certain height leaving an expensive job for property owners who wanted to build over or plant something in its place. Stumps are a costly obstruction to recovery and a reminder of life pre-fire.

The winter of 2020–2021 following the North Complex Fire levied intense storms with heavy rains, wind, and snow in the Camp Fire burn scar. Rain weakens the soil where roots barely tether a dead tree to the ground. Snow piles on fire-damaged limbs and causes them to break. I'll never forget panicked calls from Town residents begging to speed up tree removal which was prioritized for crew efficiency. "My neighbor's dead tree is about to fall onto my 9-year-old's newly rebuilt bedroom!" screamed one mother into my phone. In that case, we worked with the California Governor's Office of Emergency Services (Cal OES) to mobilize a tree crew to respond but the neighbor could not be reached for permission to enter their property so the tree remained. I recall the scene vividly: residents and uniformed crew members pacing the fenceless, checkerboard lots on their phones, the tension thick as the wind.

Day after day in Town offices and on our phones, we heard intense fear, deep frustration, trauma, and rage of fire survivors relying on the State's tree removal program for their personal safety. We stood before survivors with COVID masks on our faces, beseeching each other's eyes for understanding. Dutifully, we communicated each complaint over to the State agencies in charge, and that was all we could do.

Damaged and destroyed trees are dangerous on their own, add an ounce of wind to a community under reconstruction and they are deadly. Over time, it grew nearly impossible for me to repeat that an ineligible danger tree could not be removed no matter how threatening it was to a child, to a family, to a home. "But that makes no sense!" parents would fume at me, and they'd be right. Eligibility is defined well outside of the burn scar. Crews mobilize locally for efficiency not degree of threat. The realities of recovery make no rational sense, and my role as messenger felt paralyzing then enraging.

Counterintuitively, even if trees are deadly, they have sentimental value. After a wildfire in Oregon killed a local teenager, a conference attendee shared with me how shocked she was that the community refused to remove a stand of trees for a fuels-reduction project. She was aghast, certain that the tragic death would eliminate all barriers to reducing fire risk. My heart broke for her recovery realization. I told her it's hard to grasp dangerous levels of attachment in the face of tragedy, but it's the tragedy that reinforces the attachment. When people lose everything, they might hang on to anything that reminds them of their former lives. Even risk.

One morning, down a long street running perpendicular to the Feather River Canyon where the Camp Fire entered Paradise, I spotted a perfectly intact playhouse in a partially burned tree. Its red roof and shutters contrasted the weedy lot, and I couldn't tell if it had been in the front yard or the back.

To me, this was the saddest sight, a sign of children's lives before interruption by fire. The playhouse would never be played in again not because the tree had to come down, but because there was no one left to play. I hoped whomever owned the lot had moved their children to a joyful place where their childhoods could resume—if only until this place was home again.

In my day-to-day life, I take longs walks in my neighborhood at night, imagining what everyone is watching on TV and eating for dinner as they wind down. I am as curious about people and their choices when their neighborhoods are gone. In my Camp Fire recovery work in the Town, I could see tragedy in the shreds of lives the fire laid bare for the world to see. There is nothing more naked than a burned landscape whose shrouds and hidden spaces are gone.

As much as I sought to understand what I saw in the destruction, as I imagine disaster tourists do, I felt deeply protective of the community's private lives and wounds, and still do. I share them here only to show how the aftermath of wildfire affects our minds and hearts as much as the fire, long after media reporting on the tragedy ends.

In Butte Creek Canyon, my mom held on to the partially-burned oak tree that held the rope swing my daughters and their cousins played on prior to the Camp Fire. I took one last photo of Heidi enjoying the swing in October of 2018, weeks before it burned. The first spring after the fire a few branches leafed out and others didn't. The next year fewer leaves grew until one by one the heavy limbs splintered and dropped onto their driveway, barely missing their house. Right before the final limb fell, my mom's neighbor warned her of crackling sounds she could hear from the road.

2.1 Clipboard Fatigue

When fire damages reach a certain threshold and bring in federal and State resources, disaster recovery involves dozens of strangers conducting hundreds of tasks on private property: posting address markers on destroyed lots that otherwise couldn't be identified, household hazardous waste assessments, watershed stabilization, damage validation and documentation, site inspections, compliance checks for government contractors. After fire, if owners have signed a Right of Entry (ROE) form allowing government work on their property, they are often not present nor in the area due to displacement. For some, work of any kind on their property feels like a violation, and they coordinate each visit with each agency even if they have to travel. For others, this work is a welcome necessity bringing them one step closer to reconstruction and resumption of their lives. For many, recovery is nearly impossible to think about for the pain it brings up about the disaster. After the Camp Fire, it was not uncommon to reach a fire survivor a year or two into recovery to hear they hadn't opened their mail since the fire. If recovery isn't as traumatizing as fire, it is the most retraumatizing experience after fire. It's not uncommon in our community to hear fire survivors pledge to "go with their house" next time.[4]

Debris and tree removal in the Town of Paradise following the Camp Fire required two processes spaced years apart with two sets of permissions, deadlines, and extensions. Due to changes within State-led wildfire operations within Butte County over the past few years, we assume State agencies learned from Camp Fire operations that it's more efficient and less costly to remove debris and trees at the same time. Now, many wildfire recovery programs are consolidated, reducing multi-year processes of site destruction into a single production.[5]

During tree removal, I counted a minimum of seven agencies inspecting each lot over several months in various company vests and safety gear, some wearing name tags, most holding clipboards or tablets while walking the site looking for waterways, cultural artifacts, biological conditions, and significant archeological remnants. Crews measured the tree diameter first then the distance of the tree from the road, assessed the property in which the majority of the trunk grew (not as easy as it sounds), and evaluated the distance to a standing structure or construction site to determine if the tree could be limbed before felling. Crews included assessment, marking, traffic control, felling, large material trucks, slash trucks, clean-up, and inspectors confirming each lot for completion and compliance.

At times I felt the work was having a net positive effect. I'd refer to the Town as the largest construction site in the world and marvel at the workers, the materials, the effort, the time, the money, the creativity, the tenacity, all the things it takes to rebuild a house a thousand times. I'd see contractors on new roofs, linemen and women climbing utility poles, and underground services boring underground trenches. I'd see and feel and hear the sound of progress all around me since construction was considered essential work during the pandemic. On rainy days though, the construction workforce would disappear, the roads would clear, traffic control would shut down if they could, and progress would pause. As much as I could hardly tolerate the sadness of the best times, those gray, rainy days were the hardest.

A while into recovery it's fair to say fire survivors grow tired of clipboards on their lots, less trusting of government programs if they ever trusted them at all, and fatigued by timeline promises and delays. At a recent meeting in Magalia, community members listened to PG&E present their "Butte Rebuild" undergrounding workplan over the next two years which is a result of their responsibility for the Camp Fire. They shared color-coded maps of the Upper Ridge with anticipated start dates, the order in which they expect work to occur, and the areas not being served by the program—essentially the intact neighborhoods. Five years into recovery, residents vocalized their distrust for the best laid plans.

---

[4] Reported by Camp Fire survivors during insurance recovery operations.

[5] North Complex Fire and Park Fire Debris and Tree Removal Operations led by Cal OES and Cal Recycle occurred nearly simultaneously (2020–2021 and 2024–2025, respectively).

2.2 Long Recovery

At the County, long-term recovery staff share an office with emergency management staff. Our day-to-day is shaped by preparation, response, recovery, repeat. My mind and emotions are fastened to the fragility and complexity of our circumstances and risk, nuances I try to articulate to consultants with mixed results.

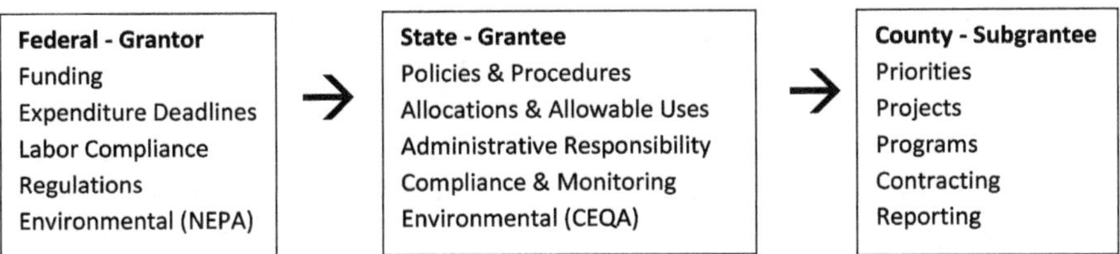

People living and working in the valley just outside of the burn scar may have no concept of what's going on within it. If it's out of sight, it might be out of mind. Or, it's up the hill but not dominating their thoughts. For most people, I'd wager, it's too heartbreaking to comprehend day in and day out. In the years following the Camp Fire, people quietly admitted while visiting me at work that it was their first trip to Paradise post-fire.

When I tell people, even my parents who were personally impacted by the Camp Fire, that we're still working on recovery at the County I have to explain why it takes so darn long. The answer itself is too long and too filled with acronyms to make sense and their eyes glaze over.

For a disaster to receive a federal declaration and therefore federal recovery assistance, it must meet a damage threshold set in the Robert T. Stafford Disaster Relief and Emergency Assistance Act (Stafford Act). The Stafford Act was signed into federal law on November 23, 1988 and sets statutory authority for most federal disaster response activities especially as they pertain to the Federal Emergency Management Agency, or FEMA.[6]

For federally declared disasters, federal grants are the largest single source of funding available for recovery. FEMA Public Assistance (PA) and Individual Assistance (IA) are the first federal funding sources granted.[7] PA partially reimburses local governments and eligible special districts for the cost of repairing public sector damages. IA goes directly to individuals to meet immediate needs. If FEMA determines an individual or agency was not eligible or actual costs are less than initially estimated after funds are disbursed, the agency will recollect the funds years after the disaster through repayment plans.[8]

---

[6] Robert T Stafford Disaster Relief and Emergency Assistance Act, Federal Emergency Management Agency, https://www.fema.gov/disaster/stafford-act.

[7] Stafford Act, Section 430, Page 76, https://www.govinfo.gov/content/pkg/COMPS-2977/pdf/COMPS-2977.pdf.

[8] Stafford Act, Section 406, Page 53, https://www.govinfo.gov/content/pkg/COMPS-2977/pdf/COMPS-2977.pdf.

Private insurance is meant to be the first-in and primary funding source for individual recovery.[9] As we've learned in Butte County, the majority of fire survivors are renters, underinsured homeowners, uninsured, or already homeless as defined by the US Department of Housing and Urban Development (HUD).[10] In a survey of Park Fire and Borel Fire survivors jointly conducted by Butte County and the California Department of Housing and Community Development (HCD) in the fall of 2024, only 9% of survey respondents assumed their insurance proceeds were adequate to reconstruct their home.[11]

For financial gaps that remain after FEMA funding is obligated, Community Development Block Grant Disaster Recovery funding (CDBG-DR) is meant to address unmet recovery needs.[12] By design, these funds are not available for several years following disaster. These grants are appropriated by Congress, allocated by the State, then prepared for application through a series of policy decisions and public comment periods. There is a distinct chain of custody for the funds that takes years to establish before the grants ever touch ground, if they do at all.

Local jurisdictions eligible for CDBG-DR funding identify recovery needs, scope projects that meet program requirements, prioritize eligible projects, apply for funds, complete due diligence, avoid choice limiting actions, receive awards, execute standard agreements, complete environmental requirements, receive notices to proceed and authority to use grant funds, begin monthly reporting on activities and expenditures, procure contractors, execute contracts, begin construction, re-scope and re-budget the projects if needed, extend the expenditure deadline if allowed, complete the projects, close out the grants, and prepare for federal and State monitoring which can last several decades. This process is completed for each grant within each tranche, with different policies and procedures for each funding round, all within an expenditure window set by the federal government. HCD worked with Butte County to seek HUD's approval to extend the Camp Fire CDBG-DR expenditure period from 2026 to 2029 to account for the many steps required to complete the projects.

For some jurisdictions impacted by disaster, the complexity of administering federal funds and the prohibitively limited ways the funds can be used are barriers to recovery. FEMA PA and CDBG-DR are reimbursement-based which means the funds must be spent first by eligible jurisdictions before they can be reimbursed. This means local governments must front 100% of the costs of recovery unless they receive a rare cash advance, in the hopes of receiving 75% or more through federal and State reimbursement depending on cost-share requirements. Finding a second source of funding to cash flow projects reimbursed by federal recovery funding may be a reason some local governments and agencies cannot engage in the FEMA Hazard

---

[9] Fellowship Interview, Clay Kerchof, CA Department of Housing and Community Development, 2024.

[10] Self-reported data gathered at Butte County Local Assistance Centers during survivor intake, most recently the Park Fire in 2024.

[11] Reported by CA Department of Housing and Community Development staff upon conclusion of the survey, 2025.

[12] Action Plans and Federal Register Notices (FRNs), CA Department of Housing and Community Development, https://www.hcd.ca.gov/funding/dr/action-plans-federal-register-notices.

Mitigation Grant Program (HMGP) or CDBG-DR.[13] In the words of Angie Mannel, Butte County Emergency Manager, "local governments are never made whole after disaster."[14]

Ironically, federal funds may be required to solve problems the disaster did not cause, or may create problems of their own. If housing is the goal, and fire survivors are ineligible for reconstruction grants, and/or the reconstruction grants are not sufficient for reconstruction, the funds go unused and individual recovery does not occur.[15] If the State then shifts those funds outside of the burn scar to pay for infill multi-family housing developments in urban areas where the fire did not occur, then they may not house survivors. I call this displacement housing as opposed to replacement housing. As of September 2025, nearly seven years after the Camp Fire, there are 2,667 multi-family housing units partially funded for construction in urban centers, while remote, rural communities that lost naturally-occurring, affordable single-family homes flounder in disaster-caused homelessness.[16]

It is important when evaluating the pace and scale of recovery to distinguish the availability of recovery resources for fires that reach the federal disaster threshold and those that do not. Fires that do not meet the damage threshold for federal declaration have fewer funding options. Local governments, tribes, and other eligible agencies and entities may apply for Fire Management Assistance Grants (FMAG) from FEMA for expenses related to response.[17] If approved, the FMAG covers 75% of eligible response expenses with the State covering the rest. As outlined in the Stafford Act, a State Governor may request a federal disaster declaration from the President of the United States when it is found that the disaster is of a severity and magnitude that State and local resources are overwhelmed.[18]

In California, local governments that declare emergencies can request funding from the California Disaster Assistance Act (CDAA) administered by Cal OES on behalf of the Governor.[19] Local governments can also request a State-led Private Property Debris Removal Program (PPDR) with an accompanying urgency ordinance to establish health and safety requirements. Following the Park Fire in Butte County in 2024, I observed this process unfold as a letter-writing campaign between the

---

[13] Fellowship interview, Taylor Nilsson, Butte County Fire Safe Council, 2025.

[14] Fellowship interview, Angie Mannel, Butte County Emergency Manager, 2025.

[15] This process occurs through Action Plan Amendments for CDBG-DR Funding, CA Department of Housing and Community Development, https://www.hcd.ca.gov/funding/dr/action-plans-federal-register-notices.

[16] Affordable Housing data compiled from Butte County jurisdictions utilizing disaster recovery funding, by the Butte County Housing Authority, September 2025.

[17] FEMA Fire Management Assistance Grants eligibility information, https://www.fema.gov/assistance/public/fire-management-assistance.

[18] Title 44 of the Federal Code of Regulations as outlined in the Stafford Act, Section 401, https://www.govinfo.gov/content/pkg/COMPS-2977/pdf/COMPS-2977.pdf.

[19] California Disaster Assistance Act, Cal OES, https://www.caloes.ca.gov/office-of-the-director/operations/recovery-directorate/recovery-operations/public-assistance/california-disaster-assistance-act/.

County and Cal OES. Establishing services and funding for recovery resembles a pen pal exchange between local and State governments.

Navigating the disaster funding world is challenging even for counties with recurring disaster experiences like Butte County. It is exponentially more daunting for survivors faced with piecing together their recovery for the first time. To this day, Camp Fire survivors bring their worn, weathered file folders to our office to hand over their insurance proceeds for debris removal costs. This is a head-scratching calculation where homeowners enrolled in a government debris removal program are required to reimburse the State with any insurance proceeds they receive for the cost of removing eligible debris, minus what they used from those proceeds to remove debris that were ineligible.[20] The County is responsible for collecting insurance proceeds from survivors to reimburse the State, then the County may be reimbursed for collection costs by FEMA. In recovery, money goes back and forth, back and forth.

During COVID, I watched survivors hunch toward our plate glass window trying to be heard through their masks. We found a chair and tucked it under the window so they'd have a place to sit for these stressful transactions. Recently, a survivor helped herself to a piece of candy saying the insurance repayment process is nearly as stressful as the fire itself.

Losing everything is surely the hardest way to be impacted by fire but it's certainly not the only way. Our community has images of wildfire burned into its collective brain. Satellite imagery from the Camp Fire is used liberally in presentations well outside of our area, often followed by the question, "can anyone name this fire?" At a conference near Vacaville, CA, an attendee raised his hand and I heard him say, "yeah, that's my house under there." I've become acutely sensitized to the re-traumatization of disaster photos on fire survivors. Colette Curtis, Recovery & Economic Development Director for the Town of Paradise shared with me they now have an unspoken rule for presentations: no flames.[21]

During long term recovery, when we talk about community safety projects like evacuation planning, the public takes notice. These are not theoretical exercises based on what-ifs; rather, these projects trigger memories of utility poles falling into roadways blocking all routes out, or photographs of burned cars lining impassable roads.

It no longer surprises me that as local government departments work with outside consultants on projects like these during recovery, clashes occur. For consultants who have no experience with evacuation in the real world, catastrophic modeling is a computer-driven progression of color-coding on a screen. It's like watching watercolor paint spread slowly across a page. For first responders, mapping evacuation routes conjures visceral memories of standing on street corners directing traffic during fire. It is inseparable from the responsibility of holding life or death in their

---

[20] Stipulated in public information shared by State agencies for the Camp Fire Debris and Hazardous Tree Removal Programs.

[21] Fellowship interview, Colette Curtis, Town of Paradise, 2024.

hands, which heightens the stakes of the project. I had the impression when facilitating a difficult conversation between consultants and County staff that one side of the table was watching a video game while the other was suffering from recurring nightmares.

I haven't sifted through ashes to find my grandmother's teacup like my mom has, then wash it and explain to everyone who admires the turquois stain that it's been "fired" for real. But I do spend large chunks of every year expecting my home to burn down. It's not if but when for many people living in Butte County and similar fire-prone areas. We spent the whole summer of 2021 watching the Dixie Fire burn neighboring communities from our front yard; in 2024 it was the Park Fire. Even when all evacuation warnings in Butte County are lifted, communities go from days to weeks to months in the presence of fire. And whereas we once breathed sighs of relief when it rained, we now fear debris flows and floods from damaged watersheds.

2.3 Fire Sprouts

When the Camp Fire burned through Butte Creek Canyon on November 8, 2018, it singed the succulents on my parents' front porch, leaving delicate leaves tinged with brown. Inexplicably, the fire burned all of their ancillary buildings but left the house intact with its rickety wooden staircase and decking untouched.

The ground all the way up to their standing home was the deep black color of coal. Trees stood with their blackened trunks and vibrant leaves along the soft gray gravel lane cutting through the adjacent meadow. Weeks after the fire, tiny shoots of green grass pushed up through the blackened earth and we saw hope again. I would kneel with my face to the ground, breathing in the scorched soil scent of wildfire, to take in the miracle of life.

Fire awakens dormant seeds, as Jim Broshears explained to me. Property owners in Paradise found things growing on their properties post-fire they'd never seen before— signs of weeds, bushes, and trees pulled out decades maybe centuries earlier.[22] Healthy fire that reinvigorates the land has been used for habitat restoration for generations, and culturally for forest management by native tribes for centuries.[23] The Camp Fire appears to have destroyed more than it awakened, but it's possible the land is simply holding its secrets for the next fire.

My parents lost their oaks, pines, and bamboo, their garage, storage shed, and carport. The fire burned the classic car my uncle lovingly restored before he passed away and left it to my stepdad. Years earlier, my husband and I had "ridden off into the sunset" in that car on our wedding day, sitting in the trundle seat while my brother and sister drove us through Bidwell Park.

My parents' insurance policy required that they itemize every single thing they lost and estimate its replacement value. This meant each spoon, plate, book, ornament,

---

[22] Information provided by Jim Broshears, Retired Fire Chief, Town of Paradise, 2020.

[23] Information provided by Jim Houtman, Butte County Fire Safe Council, 2023.

knick-knack, things they couldn't even remember they had, stored away in boxes for their grandkids.

My mom cried for the fact that her house with her day-to-day stuff remained intact, while family treasures kept by both of her parents until they died burned in their storage shed. She is a keeper of things and meticulous in her keeping. The possessions she lost represented the love of her parents, memories of her childhood, her family heritage, and her emotional wholeness. I've heard many fire survivors say losing the only tangible items left by loved ones is like losing them all over again.

During the months my parents prepared their insurance claim, I listened to everything they lost. I read through the list, edited, advised, and cried only when I learned the Christmas stocking made for me when I was a baby had burned. Otherwise, I was grateful they didn't lose more, though their property continued to change as debris was removed, soil was excavated, and grounds were remediated. One by one their beautiful trees fell from the slow death of fire damage and stress. I've learned to take mental snapshots of recovery over time, to remember all the things we've yet to lose but will.

To this day, when my parents' neighbors see a tree down in Butte Creek Canyon, they roll up in their tractors to chainsaw and push slash into burn piles. When folks are driving too fast, they dig and pack their own speed bumps. Maintaining their environment is a way of life. Whereas my parents are relatively new to rural living having bought their house in the canyon just one year before the Camp Fire, their neighbors regularly lend a skilled hand and heavy equipment.

2.4   Resilience Spectrum

I often think about what resilience means in the face of fire, and if it is achievable and lasting. I think about my own resilience, my community's resilience, the planet's resilience. My youngest daughter loves to talk about how nature will overtake society within 50 years if cities are abandoned by humans. She loved seeing pictures of coyotes roaming the streets of San Francisco during COVID.

Resilience is the period everyone puts on the end of each sentence that starts with disaster: disaster, response, recovery, resilience. It flows naturally and linearly to a finite point, making people feel better about disaster if resilience is the result. But is it?

Community resilience seems partially attainable when motivation is high post-fire, but there are limitations due to cost, scale, and incentive. I sit on insurance webinars day after day learning how to Teflon our world: gravel our yards, remove the trees, harden the homes, repeat the recurring cycle of enforcing the weed ordinance because weeds won't police themselves. If we do everything we've learned to protect the built environment, is it conscionable to say those things will save all lives and property at risk during catastrophic fire?

As we know here, the universal 'we' leaves out a lot of households like those on large wooded lots without adequate funds for fuels reduction, and those who won't sacrifice the beauty of their land to reduce fire risk.[24] On a webinar, I heard the Fire Administrator for Napa County say much of his work is helping people grieve the nature they moved there to enjoy, so they can remove it to reduce the risk they'll lose it and more to fire.[25]

With enough government subsidies and willing property owners, vulnerable communities can accomplish a percentage of these recommendations in fire-prone areas, but without 100% participation does the risk decrease enough to justify the cost? Feasibility depends upon willingness, resources, and maintenance, and lies somewhere in the middle of the resilience spectrum. Resilience may be just as costly as recovery, and it is not the same thing.

In the aftermath of the Camp Fire, retired Fire Chief Jim Broshears, told me we plan for the worst we can handle not the worst that can happen.[26] I've never forgotten those words and have come to experience, over and over again, the worst continuing to surprise us because we fail to think in those terms. We plan for all the ways we can respond as humans, but fall short of planning for unimaginable circumstances we simply cannot foresee. It's been said that catastrophic fires like the Camp Fire make national news because the destruction demonstrates the limitations of human response. It may sound like we're talking about fire, but we're really evaluating ourselves. Everybody knows humans can't stop tornadoes, so we use the word tornado to describe out of control fires as a means of acknowledging our humanity.

As Jim explained to me, evacuation zone mapping in the Town of Paradise, pre-Camp Fire, was meant to establish a grid for gradual, orderly evacuation, not to evacuate the entire ridge at once. He told me this didn't mean the planning effort was wasted, it just wasn't suited for the very worst that could happen.

Jim taught me that evacuation mapping is useful for 99% of fires that occur, with the Camp Fire being the 1% exception. When fire comes from one direction it is conceivable evacuation zones can be called in gradual, sequential order. When fire shoots itself across town and starts burning inward from all sides gridlocking traffic, people end up lying in parking lots watching fire blast over their heads at 100 miles per hour.[27]

---

[24] Information provided by Taylor Nilsson, Butte County Fire Safe Council, on a tour of fuels reduction projects in July 2025.

[25] Information provided on a webinar hosted by United Policyholders, 2024.

[26] Information provided by Jim Broshears, Retired Fire Chief, Town of Paradise, during preparation of the Camp Fire After Action and Corrective Action Report, 2020–2021.

[27] Information shared by Jim Broshears, Retired Fire Chief, Town of Paradise.

2.5 Petri Dish

Working with emergency management I see how planning and mitigation must overtake daily operations in order to reduce risk. In rural local governments where capacity is a concern during the bluest of blue-sky days, and several critical positions are held by one person, the inverse is often true. Intact communities may be unwilling or unable to stop the flow of the day-to-day with enough time to prepare for the worst they can handle, let alone the worst that can happen.

Butte County is a bit of a wildfire petri dish. Researchers regularly call to see if what we're doing is working. It's encouraging to hear from people in academia trying to solve problems, and we give interviews when they're in the best interest of the County, but the lessons often get diluted by competing priorities. Wildfire is important to us, but is it important for everyone else…yet?

One of the traditions I started with Adela when we lived in Paradise was blowing bubbles from the front porch when it rained. It was thrilling to hold our breaths as the bubbles floated out into the rain and popped one-by-one. Occasionally a bubble would make it all the way across the lawn and over the box hedge above the gravel road. That's the hope of being studied. I give every effort to every bubble I send out into the world, hoping it reaches far enough to make a difference, to beat the odds, to convey the fragile lessons it has endured treachery to teach. This book is my bubble and I'm holding my breath.

Heartbroken volunteers feeling compelled to help after the Camp Fire reached out to me when I was Disaster Recovery Director for the Town. Instead of doing their own research and offering assistance, they asked me to find a need they could meet, then help with meeting it. In my experience, this is a well-intended ask but not an efficient use of a recovery professional's time let alone emotional capacity. In short order, I went from feeling grateful for these calls to having no space in my heart or brain to assist people trying to assist. Thankfully, we had a number of hard-working non-profits in the community coordinating volunteer work, and I was happy to refer them over.

Volunteers do incredible work in disaster recovery. They raise funds, navigate the local permitting process, and build homes with and for survivors. There is some volunteer-driven recovery work occurring here in Butte County but not to the scale it occurs in urban areas or after hurricanes and floods.

I interviewed Dan Dunmoyer, CEO of the California Building Industry Association, for the fellowship. His answer to rapid wildfire recovery is streamlining master planned developments.[28] This makes perfect sense in urban settings where small, adjacent lots can be purchased and developed at scale using one or two master-planned designs with the same contractors. I understand from a conversation with William Siembieda, Professor of Planning at CalPoly, that after a fire in Southern California, a county approved building plans on dozens of contiguous lots under one permit, cutting down on paperwork and time for developers.[29]

---

[28] Fellowship interview, Dan Dunmoyer, California Builders Association 2024.

[29] Conversation with William Siembieda, Professor of Planning, CalPoly, 2024.

In my observation, and what I try to share with every researcher who calls, is recovery in an urban fire footprint is very different from recovery in a rural fire footprint. For the most part, with connections to municipal services like water and sewer, recovery in an urban fire footprint can occur in place. New homes can be reconstructed where structural debris has been removed.

In a rural fire footprint where lot sizes might be an acre or more and dependent upon septic and wells that are likely also damaged or destroyed, reconstruction processes are unique and unrepeatable. Amy Rohrer with Valley Contractors Exchange explained that contractors are hard pressed to rebuild more than one lot at a time when there are miles of damaged roadways in between projects. In some cases, rebuilds happen well outside of the burn scar within urban spheres where development is more feasible. This changes the ratio of urban homes to rural homes, weighting heavily against the latter.[30]

---

[30] Fellowship interview, Amy Rohrer, Valley Contractors Exchange, 2024.

Chapter 3
Grief

A few months ago, I was having my blood drawn in a lab when a young boy started crying a few stalls over. He sat in his stroller at counter-height while his dad used his arms to brace his body for the phlebotomist. The boy was terrified of needles or the sight of blood, both perhaps, but he was too young to explain. His pregnant mom had scooted behind my chair on her way to the lobby, likely hoping her absence would speed up the process.

The boy howled. He cried and begged and pleaded at the top of his lungs. As he resisted, I heard the tension in his dad's voice, soothing and apologetic but stern. "I'm sorry, buddy, but we need to do this." A few minutes into the session the boy fell into a pattern of wailing, "I want to go home! I want to go home! I want to go home! I want to go home!" in a steady unbroken rhythm. My nerves were fully activated by this time, and my body tensed from the trouble my phlebotomist was having finding my vein.

Finally, blood drawn or not, the dad wheeled the boy's stroller behind my chair toward the exit. As they went, I heard, "I want to go home!" at top volume. My eyes filled with tears for the young boy's distress and the worry his parents must have felt if this was diagnostic. He was saying, "I'm scared and I want to go to my safest place in the whole world." He wasn't crying for toys or snacks, the simple comforts and conveniences parents usually pack in strollers to sooth toddlers during situations like these. His cries about home were coming from the very center of his fear. I realized that as young as he was, his definition of home was the same as mine, as my parents', as anyone's I imagine, and I thought about all the ways in which I've heard this cry since the Camp Fire.

Now I hear, "I want to go home," differently, and I hear it everywhere. Sometimes it's said with a bit of urgency, sometimes as an order, but a lot of time as a calm expression of retreat. It's the sound of a father calling for his kids at the park as the sun goes down. It's the excitement of my co-worker as a long weekend approaches. It's the answer to many polite social exchanges about what's next in the day. But whereas a terrified young child can shout it freely at the top of his lungs in a hospital,

K. Simmons, *Three Fire Mountains*, https://doi.org/10.1007/978-3-032-17343-0_3

adults—fire survivors who have lost their homes—have to express it in a hundred different mostly silent ways.

Home is the physical space we reference most. It is the place we commonly map to and from on our phones. When we lose a home to fire or flood or another calamity, our roof and our walls are gone but so is our safety. For many fire survivors, safety isn't restored until home is restored and that's not always accomplished with a house. My dear friend who lost her home in Paradise has moved three times in the past five years, each time into enviable, beautiful houses. Just last year she wept in my backyard out of homesickness for the scent of the pines and her sense of community. This was her saying to me, "I want to go home." She lost her weekend retreat in the Dixie Fire, too, so any sense of familiar respite from her unwelcome urban environment is gone.

I have learned home is not a one-to-one replacement after disaster. Even those who did not lose their houses in the fire lost the idea of "home," and the safety that comes from taking that for granted. Even for me, living a mile from the edge of the Camp Fire burn scar, my faith in home as a permanent place I can rely on is shattered by what I have seen.

In my parents' case, even though they had a roof over their heads each night, living without their home for a period of time following the Camp Fire quickly turned into a life-threatening hospitalization. For other survivors, it may show up as prolonged depression, anxiety, post-traumatic stress disorder, or other chronic health conditions.[1] From the courses I've taken on resiliency, I see how predictably my life has shaped itself around a fire-driven hub of thoughts and feelings. With some training, I no longer see my reactions as unique; I am a posterchild for post-traumatic stress from the way fire has reshaped my existence.[2]

The County's Department of Employment and Social Services hosted a meeting in late 2023 with guest speakers who'd experienced homelessness. One speaker said the hardest part of homelessness was not knowing where to take herself during the day. She literally did not know what to do with her physical body. She would keep moving so there was always a destination because there was never a destination. She wished for a home or an office or a park that would allow her a sense of arrival and rest, a single place that was predictable without feeling like she had to move again. As she wept in front of us, I heard what she was saying: I want to go home.

Before hearing her story, I'd never considered that a body can be a burden when it has no place to go, or what a crushing responsibility that is. Indeed, we see fire survivors suffering the same physical and mental afflictions as people experiencing homelessness. An inability to be as pure and honest as that little boy obscures what so many people are crying inside.

---

[1] From anecdotes reported to me by fire survivors, case managers, and recovery professionals over several years following the Camp Fire.

[2] Personal observation after participating in resilience training provided by Butte Glenn Community College—The Training Place between 2023 and 2024.

Oakland Firestorm, October 19–20, 1991

When I was 14 years old a fire came within miles of my childhood home in the East Bay Area. The Oakland Firestorm made international news because within 1,520 compact urban acres it killed 25 people, injured 150, and destroyed 3,810 structures.[3]

The day the fire broke out I remember my parents watching the news and talking to our neighbors in the cul-de-sac, eyeing the sky and trying to get a sense of the magnitude. As a family and as a community we were not attuned to reading the size and color of smoke plumes to know what was burning, where, and how quickly. Black smoke billowing out from all sides is a structure fire. Thin brown smoke gradually dissipating into the blue sky signals a pile burn. Cauliflower-like thunderheads suggest a massive fire is on a run in the mountains.

The night of the Oakland Hills fire we could see the fire glow in the distance like a large flickering candle. A top concern among media broadcasters was the Claremont Hotel, an historic ivory castle perched on a hillside below the UC Berkeley campus, visible from miles around. The plight of the hotel and the possibility the fire would move through Orinda into Moraga kept us all in suspense, but I don't remember being afraid of what I later learned fire could do.

The destruction sent shock waves through the community that lasted for years, maybe to this day, much like the 1989 Loma Prieta Earthquake a few years prior. Now I understand as a disaster recovery professional, particularly from the guidance of Jim Broshears, how many lessons were learned from that response and recovery to update broadly-used fire policies and procedures. Back then, however, I don't recall settling into a mindset of fire-adapted living in the Bay Area. The Claremont Hotel survived and on the very next ridge it seemed the community doubled down on its lush oak forests obscuring the homes that weren't destroyed. I don't remember a single conversation about defensible space though my parents owned nearly an acre of grassy, wooded hillside. It was as if the community took a collective sigh, shook off the fire as one-and-done, and pulled its densely populated rolling hills lifestyle right over its eyes.

The swath of devastation on the west side of the Caldecott Tunnel was clearly visible from across the Bay which is the view shed those homes were built to enjoy. The burn scar was dwarfed by the great sparkling city beyond and the twinkly lights from homes packed left, right, and center. This was the Bay Area after all, not a community in a rural, remote mountainside tucked away from the public eye. For those living east of the now craggy, tree-less Oakland hills, it seemed the trials and tribulations of life in the Bay Area resumed and everyone returned to vaguely fearing "The [next] Big One."

My experience with the Oakland Firestorm probably explains why people living just miles from the Camp Fire burn scar may be tuned out, and why those who've never seen the North Complex Fire burn scar can't imagine post-fire living conditions. On a call with our hazard mitigation consultant, Jeanine Foster, who, coincidentally,

---

[3] Ewell (1995).

graduated from the same Bay Area high school a few years before I did, remembers the fire differently. Her dad was a dentist in Oakland who produced dental records to identify fire victims. Her friend's sister died in the fire. She remembers the aftermath vividly to this day, her experience much closer and more personal than mine.[4]

This goes to show how differently people can react to the same fire, how proximity to death and destruction matters even within a few miles, and how age and experience can shape perception and memory. I didn't run for my life from the Oakland Firestorm so though it is seared in my mind, and I watched the reconstruction of those homes and multi-family housing units for the next decade, it didn't change the way I lived from one day to the next. It was not a demarcating trauma from my childhood separating before from after. It was a very tragic incident but for me it wasn't the worst.

3.1 Bleeding Heart

Of all the Camp Fire maps we look at, the sectional damaged and destroyed map in CAL FIRE's Damage Inspection Report is the hardest for me to see.[5] A tiny map of the burn scar in the lower left shows the section of the fire enlarged on each page, and there are many pages. One box at a time, one page at a time, tiny red cubes dot the landscape, each structure accounted for, all the homes destroyed in the fire. Page after page of cul-de-sacs, lanes, roads, avenues, streets of an entire town and surrounding areas lost save a few standing structures. When I first started my job in Paradise, I had a large printed map of the burn scar on my wall with each red cube blending into the next. "The bleeding heart of Paradise," Marc Mattox, Director of Public Works for the Town of Paradise, said, as if it was an open wound.[6] And it was.

I didn't know then what I know now which is having disaster imagery in front of my eyes full-time, and in front of the eyes of everyone who came into my office, was damaging in and of itself. I now dole out those images infrequently for myself and others, and think carefully about the impact on those I am sharing them with. I don't look at fire maps unless I have to, often only to orient a consultant or contractor to the area I'm talking about when I refer to the burn scar. "Burn scar?" one consultant working across rural California said to me, "I've never heard that term before."[7] I was shocked and angry as I often am when I have to start from scratch. "See, there was this fire…" I start, with my nostrils flared. Even now, the story is impossible to retell without feeling the rush of it on my skin.

---

[4] Information shared during the preparation of the Butte County Local Hazard Mitigation Plan (2024–2025) by Jeanine Foster, Foster Morrison Consulting.

[5] Camp Incident Damage Inspection Report, CABTU 016737, November 26, 2018, Submitted by Nick Wallingford, Damage Inspection Manager, https://www.nist.gov/system/files/documents/2020/11/16/2018 Camp Incident DINS Final Report.pdf.

[6] Spoken by Marc Mattox, Director of Public Works, Town of Paradise, 2020.

[7] Spoken by a consultant during the development of a plan for Butte County.

When I worked in Paradise, I would drive up the hill from Chico early enough to explore the devastation before clocking in. Now I see that as picking at a scab that would literally never heal from my constant interference. In truth, I would probably not describe it as picking, I would say I was excavating the disaster and its aftermath aggressively, as if frantically digging in the dirt with my bare hands to find something I couldn't live without.

Figuratively, I would dig and dig and dig, searching for meaning, looking for the truth, the reason. TELL ME WHY, I seemed to be screaming silently at the disaster from my car. Backlit by the sunrise, tall stands of skeletal trees glowed like the backdrop of a tragic stage production I watched from the front row. I now see this insatiable desire to comprehend the inexplicable as toxic and damaging without proper support. I ate up the devastation with no way to metabolize it.

One day I pulled over to look at a row of multi-colored mailboxes down a shared dirt path leading to nowhere. Nine boxes serving zero houses. I wondered who once collected mail from those boxes and where they had gone. I stopped and took pictures of mid-sized maples frozen into giant whisks churning the air, and imagined the moment the tree burned into that shape, fire curving its branches. Occasionally I'd see daffodils and tulips sprouting along the driveway of a vacant lot and wonder if the bulbs had been there before the fire or if they were a beautiful addition left by a heartbroken survivor. I was searching for answers and getting secondary trauma in return.

Looking back, I was a mid-career wife and mom trying to do good things for my community from the depths of my new disaster recovery profession. I became a dogged investigator, partially motivated by thinking if I understood the disaster, I could understand the recovery. And maybe I could understand why I was there. Why I chose this.

I've always had a deep social curiosity that is not harmful on its surface, but this experience taught me that if it's left wholly unchecked it can nearly undo me. Now I protect myself if I can by moderating my exposure to tragic images, knowing how susceptible I am to memorizing and internalizing them.

I got out of my car one morning to study boulders held tightly in an upturned root ball the size of my car. Until then, I hadn't known standing trees were gripping rocks beneath the surface with their roots, holding on to them as they held themselves up. I didn't know tree roots would take those boulders with them when they were pulled out of the ground. Another root ball held the white PVC circuitry of a sprinkler system, now gnarled and gnawed by twisty roots. I learned things about healthy, standing trees only fire-damaged trees could reveal.

3.2 Binary Fire

By 2020, Paradise was dotted with standing homes, soldiers fully exposed and on guard. Fire is relatively binary, destroying nothing or everything. Even if fire destroys a home only partially it is often considered a total loss for the work required to remediate and repair it. From studying wildfire so intently, destruction became my baseline, and I started to see everything outside of the burn scar as intact. Whereas once a house was just a house, now if it hadn't burned down, it was an intact house. Whereas I once saw a community as just a community, a community that hadn't burned down was intact. Intactness was now 'other' to the destroyed norm (Fig. 3.1).

Down one long cul-de-sac where there was nothing left but boulders and root balls and trees falling weekly, there was a beautifully intact home with purple maples and gentle white blooms sparkling in the morning sun. Seasons don't come to newly destroyed communities like they do to intact communities where there are leaves to change. One block could appear a barren wintery landscape while another is in full-blown fall.

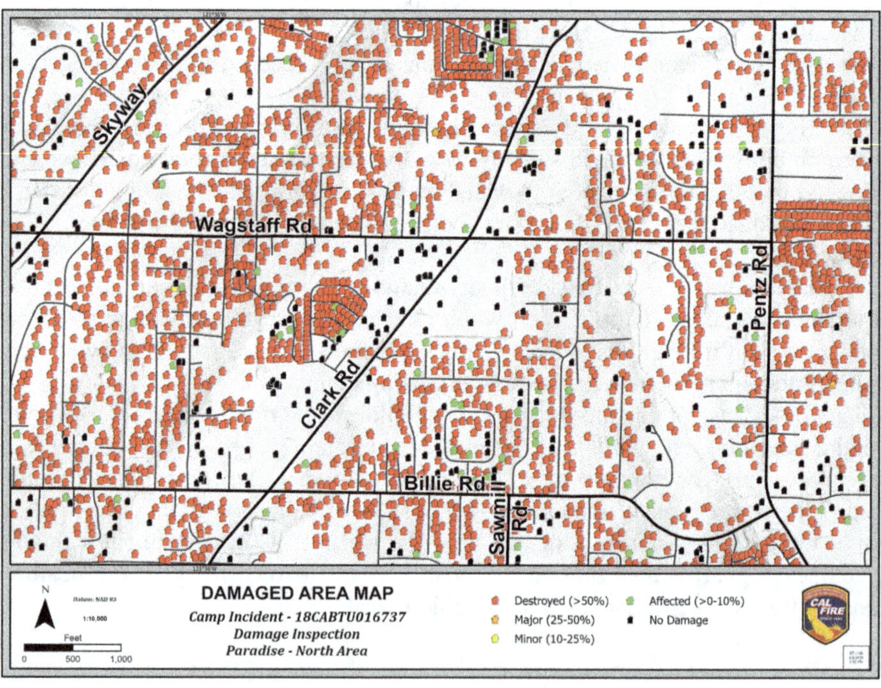

Fig. 3.1 Cross-section of damaged and destroyed structures from the Camp Fire Damaged and Destroyed Structures Report, map produced by CAL FIRE (Originally published by NIST)

A few times I found myself pulling up Google Maps when familiarizing a consultant with the town pre-fire because while some mapping had been updated with new video, some had not. There was one neighborhood along Pentz Road where scanning south on Google Maps you could see the bare brown earth of destruction, and scanning north you were transported back into a time of delightfully landscaped neighborhoods with dappled shade and vibrant colors. It's hard to appreciate how colorful neighborhoods are until they're no longer there. Even today, some maps remain a perfectly preserved "before" picture of Paradise with leafy branches curving a canopy over the street. At any given time, there's probably little continuity between what's on Google Earth and what's on the ground given how quickly the town's landscape is changing. Over time, I imagine the many faces of Paradise will merge into one as its pre- and post-fire states disappear.

On those morning drives, I would set my intention to celebrate the rebuild in an attempt to stabilize my outlook swinging wildly from hope to despair. I remember trying to explain the two sides of recovery to a friend who said, "but you felt great yesterday—what happened?" Construction starts early in the day so nail guns were popping at the crack of dawn, workers were waving each other over, pointing, guiding deliveries of lumber, rubbing their faces, drinking their coffee. I would see life carrying on all around me and try to lock into a positive head space so I could push through the recovery bureaucracy and devastating stories at the office.

From my car, I watched squirrels scurrying around freely and deer roaming unobstructed by fences, signs that wildlife was returning. One day I saw a handmade sign taped to a street pole looking for a lost turtle. Surely a nomadic turtle was evidence the town was returning.

I could see the restoration with my eyes and hear it with my ears, a full sensory experience in Paradise. The coursing energy of time and activity being reinvested was visceral and I was grateful for it. I felt called to be part of that rebounding circulatory system bringing energy to the recovery, but did not have the language nor tools to care for myself in the process.

I tried mightily to manufacture enough well-being to keep going. I posted my hopes and dreams for the town on social media. I reflected on my journey back to Paradise and recruited new staff using phrases that had worked on me like "once in a lifetime opportunity." I projected the currency of hope inwardly and outwardly in all the ways I thought I should.

My mom bought me purple embroidered cowboy boots I wore proudly while whipping my Prius around town, eventually ripping the plastic undercarriage off the vehicle during low clearance excursions down private gravel lanes. On my best days my enthusiasm made for a breezy ride, on my worst days grief nearly paralyzed me. I wonder now if an outside support system stabilizing the highs and lows of recovery would have sustained me in the job. To achieve that—because it might have been available had I asked for it—I would've had to have known and accepted my own limitations. But that's not what a "hero" does; a hero goes all in and burns out.

Because I learned on the job, I immersed myself in the language and structure of emergency response, observed the way hazardous tree removal calls between agencies ran like incident command briefings with situational status reports. I grasped the finer points of eligibility, the expenditure timelines of different recovery grants, and made spreadsheets showing the who, why, how, when, and where of each project. I studied the recovery plan, watched videos of its creation, and spoke at community meetings on recovery progress. Weekly, I appeared on the local news before the recovery stats leveled off and COVID took over.[8]

The Town's financial consultant called COVID "a blip" to the "cliff" the community had already experienced as a result of the Camp Fire.[9] We moved air scrubbers into our workspaces, wore masks, rolled down the windows when we drove together. While the rest of the world worked through the complications of COVID, we added COVID to the complications of working through a disaster. How the world's worst pandemic in a century was called a blip shows you the size of the cliff we were dealing with.

I remember sitting in the parking lot of Town Hall one day, my Prius pressed between heavy-duty trucks like a missing sock in a load of sheets. I stared out at the wispy trees made of charcoal and soot and couldn't feel anything alive inside of me anymore. I had synced up to suffering of the town and couldn't feel the progress of incremental recovery against the tidal wave of disaster. I lasted just one year working for the Town of Paradise and spent the next several years wondering why.

3.3 Life in Paradise

I moved to Paradise in 2008 and stayed until 2011. I found an ad for a small rental house in the local newspaper, my little blue house on the hill, and settled in with my daughter. Her dad and I separated during that time and he stayed in the Bay Area where he continued working for the City of Berkeley until eventually relocating to Chico. My little blue house had one bedroom, a tiny porch office without insulation, and a closet-sized room off the living space that I decorated especially for Adela. Her canopy bed that once belonged to my sister squeezed neatly into one side of the room without an inch to spare on either side, though Adela slept on the floor in her "pillow bed" most nights.

Neatly groomed box hedges edged the little blue house in the front yard, and a chain link fence separated the back yard from the house on the flag lot. A beautifully symmetrical shade tree grew in the center of the front lawn. The kitchen floor was coated in creaky yellow linoleum that made for a perfect dancing surface for our

---

[8] Action News Now, Chico, California, 2020–2021.

[9] Observed from an economist working with the Town of Paradise to determine a new fiscal model for wildfire recovery.

impromptu "dance parties" in Adela's tutus and my old prom dresses. The front room had just enough space for a couch and the "whining chair" I labeled in a glittery sign for my expressive toddler.

My bedroom was on the other end of the miniature house next to the only bathroom I decorated with a rainbow shower curtain and pink polka-dotted towels. The wall between the kitchen and the living room had a two-sided space heater. On cold mornings, Adela would stand on the carpeted side in the living room and I would stand on the linoleum side in the kitchen while we got dressed, talking to each other through the hissing slats while trying not to burn our noses.

In the backyard stood a drying rack on a swivel for clothes, and a shed that held a washer and dryer. Behind both houses toward the main road stood dozens of tall pine trees creating a forested shield. Just down the lane was a beautiful, historic brown church on Scottwood we'd run to on winter mornings leaving footprints in the snow.

Adela's preschool was a short drive up Skyway and after she "graduated" in her light blue gown she went to Paradise Elementary School right down Pearson from our house. Even though we were minutes from school, I learned right away that to get her to class on time she needed at least two hours to wake up and get ready. Her kindergarten classroom had huge leaded glass windows that looked onto the front of the school, a defining old-fashioned feature of the building.

Her kindergarten teacher loved Curious George and wore cute, yellow t-shirts featuring the cartoon monkey. Curious George himself was available to go home with the kids on weekends, with his little suitcase and journal. On our special weekend, Adela took him to a grocery store in Chico.

Recess was held under the trees in a small playground with lines painted for foot-powered cars. I tried to volunteer at least once a week and on field trips.

When the weather was nice Adela and I would walk to school along the rocky street. There were no sidewalks coming in and out of our little neighborhood and it was noisy with cars rushing by in the morning. I loved the idea of walking but it wasn't practical even though we could get between home and school in 15 minutes. We drove most days so I could race to work after walking her to class.

After the Camp Fire, Paradise Elementary School was gone but a bus yard filled with rows of yellow buses stood at the back, clear as day from Pearson. Seeing buses without schools, without kids, without a place to go during COVID, brought me to tears every time.

I've never met anyone happier than Adela's kindergarten teacher, or perhaps more perfectly suited to her profession. At the time I was in my early thirties, getting a divorce, figuring out the rhythms of single parenthood, highly ambitious but in the non-profit sector which paid very little. Adela's teacher shared her husband's name as did her three sons. "Mom," my kindergartener exclaimed in disbelief after school one day, "you'll never believe it! My teacher's whole family has the same last name!" This would not be the case in our family, particularly after I remarried years later.

Paradise was a place of deep rest for me even with the high stress of separation and long drives to share custody. The air was clear and clean, the people were friendly, and I felt solid in my adulthood for the first time, supported by a group of incredible women who had also been around the block a few times. I'd never heard "you're welcome" as many times as I did in Paradise. In the Bay Area, "thanks" is the end of every social exchange. After almost two decades in Butte County I still feel sorry to delay people with, "yes, of course."

Living in my little blue house, I was convinced I had a fairy godmother—probably yard maintenance my landlord hired when he saw how little I was capable of doing. My yards were cleared and fresh when I got home from work. On hot summer evenings Adela loved to water the tree wearing her plastic pearls, standing under the spray to cool off. After I left the Paradise Chamber and went to work in Chico I'd drive up and down Skyway to and from work, watching the sun rise in the morning and set in the evening—the biggest sky I'd ever seen.

The drive up to Paradise took my breath away the very first time I visited. Skyway follows the top of a ridge flanked by Butte Creek Canyon on the north side and the Feather River Canyon to the south. The road extends into the foothills and eventually into the mountains. It is the canyons on either side of the town that are most fire prone, but I didn't know that on my first visit which, as it turns out, was to attend the town's iconic Johnny Appleseed Days.

I fell in love with Paradise at first sight. It was a fairytale community of good people. There were daily conveniences like banks and grocery stores and fast-food places like Foster's Freeze that had long since left the Bay Area, but the town felt quaint and small. The neighborhoods were a mix of lovely homes, rotting homes, and everything in between. My search for a rental house had me touring unsightly, smelly places one minute and gorgeously landscaped ranchettes with circle driveways the next. The little blue house was perfect.

3.4 Paradise Ladies

At the Paradise Chamber my visitor center volunteers were in their sixties and seventies. They coached me on life choices and invited me over for tea and Mary Kay make-up sessions. Well, they invited Adela over and I'd go along for the ride, admiring their lovely homes and grown-up belongings.

The Chamber Board of Directors was chock full of women who were running businesses and mini empires on the hill. They immediately circled me in the tightest, warmest hug. They could see my vulnerability which, to this day, probably sits on my sleeve like a squawking parrot, but they never made me feel too young or inexperienced for their respect.

Adela was a tiny executive with curly hair who spoke to these ladies as her peers. They had deep patience for her and for me and my 11 p.m. calls when my little executive would debate the benefits of sleep. They were my lifeline, my best friends, in many ways the loves of my life.

After the fire, I felt like I abandoned them after they had given me so much. Their pain was infinitely larger than anything I could hold as their homes smoldered and their lovely trees and porches and cars and pools and gardens sunk into pits of ash. Despite that, as women they rose like the beautiful phoenixes they are. They withstood years of RV living while they cleaned out debris and trees, settled their insurance claims, negotiated their Fire Victim Trust settlements, applied for rebuild permits, endured labor and material delays, and piece by piece rebuilt their homes and their lives. As I write this, five years after the Camp Fire, one has just settled into her new home in Paradise.

When I first took the job at the Town in May of 2020, I reconnected with Melissa Schuster who took me to her property and through the lot of RVs where friends and family stayed. We sat in her quaint little chapel which survived the fire, where Adela and I once decorated Christmas trees, and she treated me to delicious food. She was sturdy and strong and I wept for losses I hadn't yet processed, right there in the epicenter of hers.

She was thrilled and complimentary that I was working for the Town, but I would splinter and crumble beneath the weight of what I thought I could do. In January of 2024, she and I posed for a picture together at a sparkly event we used to attend annually long ago. Her hair is now silver and her clothes are fashionable as ever, and her husband embraced me in a friendly squeeze just like old times.

3.5   The Skunks

The story of my life in Paradise is not complete without the skunks. Adela was young when we lived in the little blue house and she took lots of baths. She'd be in there splashing happily and I'd be in the kitchen talking back and forth with her while I made dinner. The bathroom was simple but pretty and I was so proud of my pink towels and rainbow shower curtain. Pink was the color of our lives in Paradise.

One day I heard scratching while in the bathroom. Eerily, it sounded like it was coming from under the tub. But how could that be? Had someone crawled under the house and was now dragging their nails along the underside of the bathtub? It would start and stop, scratch, pause, scratch. We had guinea pigs at the time in a big cage on the kitchen floor, and the noise didn't sound entirely unlike Princess and Sparkle dragging their nails along the linoleum.

The scratching continued until one night the smell of skunk bloomed rich and deadly inside the house. It woke me up and clouded my senses, filling my mouth and nose with stink. Something propelled me outside, perhaps suffocation, and when I arrived on the porch the night air smelled piney fresh. Inside: toxic. Outside: cool and clean.

Skunks spraying *inside* the house?

As we'd soon learn, the scratching was the sound of male skunks clawing their way under the bathroom floor and into the nest of a female bedded inside the walls. Animal control set up several cages and over the course of a week and a half captured fifteen skunks and three raccoons. Rodney, my boyfriend now husband, fed the jailed rodents peanut butter and apples, tossing the food toward the cage then running out of the spray zone. Adela and I would wake up in the morning and she'd look outside and call out the caged number. "Mom, 4 today!" she shouted one morning before kindergarten.

During that memorable season she and I would arrive at school and work, respectively, drenched in skunk smell, our clothes moist with stink, our food emanating a rancid odor. It was embarrassing but more exhausting than anything, at least for me. Animal control explained the nesting female was in heat and male skunks were fighting over her, spraying each other, the walls, the air, everything, during their skirmishes. Toward the end of the trapping, a female skunk was apprehended but animal control didn't think she was the one. "Watch out," he said as he removed the traps at my request, "once a female goes into heat somewhere, male skunks return to the same place every year to find her." I moved to Chico shortly thereafter to be closer to my office and away from the skunks.

Paradise was the forest living I'd signed up for. It was snowy and quiet and filled with wildlife and the activities of a tight-knit community with clubs, events, and traditions. The parks were new and clean, the bike path was a perfect distance from home, the moms of Adela's classmates were thoughtful and earnest, and some welcomed me into their homes.

Laura and I took turns having playdates for our kids at her large, lovely home, then at my quirky rental. She loved the whining chair. Our favorite evenings were poolside at the gym where we'd swim then sit in the spa looking up at the trees and the stars. It was magical. That's where we were the night before she was tragically struck by her own car while reaching out to pick up something, her kids strapped into their car seats in the back. I held my breath for those few days before she died in the hospital down in Chico, her kids just 4 and 6. Rodney went to her funeral with me and we sat in dazed astonishment and grief with hundreds of familiar faces.

Paradise is where I found my love of riding bikes. It is hilly and I was a strong climber as I discovered. Racing downhill on two wheels was a thrill I'd never experienced. It was the start of many, many years of fast, exciting road riding before I finally sold my road bike and stuck with mountain biking which I still do to this day. It was coursing down the mountainside with the wind pushing into my helmet and through my hair that I found an exhilaration I hadn't felt in a long time.

Before my dad died, he liked to tell the story of watching me fly down my grandma's driveway in Novato on my big wheel. I'd kick up my little kid feet and rip down Holly Lane in a shuddery plastic flash. "I called you speed demon," he'd chuckle about my youthful bravery. That's what I found again in Paradise, a place to kick up my bigger kid feet, wild and free.

3.6 The Skyway

After I moved out of Paradise, work and home were both in Chico but I sometimes found myself halfway up Skyway after a long day. Skyway doesn't have many places to turn around so sometimes I had to go all the way up before coming all the way down. I'd shake my head, frustrated at the wasted time, but it was Paradise calling me back. Eight miles up and eight miles down, big skies the whole way. Better, clearer light than anywhere I'd ever been.

From Skyway, when the rice fields in the valley are flooded they sparkle like a glass mosaic. When the orchards are in bloom, the ground is coated with a pink dusting of cotton candy. A good Samaritan leaves jugs of water on the uphill side of the road during hot summer months so drivers can cool their overheating engines. A lookout fence on the north side is lined with locks engraved and painted with lovers' initials. After the fire, Skyway has far fewer trees, many fallen or limbed and topped, but as oaks do in fire, some burst out with new growth, and saplings dot the landscape.

During the Camp Fire, the Sheriff opened both sides of Skyway to one-way evacuation traffic.[10] This is called contraflow, cars on both sides of the road streaming downhill in the same direction. Disconnected road segments on the western slope of town left people stranded in their cars with no way out. Skyway was the most reliable road out during Camp Fire evacuation and where it once slimmed down to two lanes through town as a traffic calming measure, it has since been opened up, ready for the next time.[11]

Imagine heading downhill away from the fire only to reach a gravel road that narrows into a dead end. Connecting road segments is priority work of the recovery staff at the Town, the focus of federal earmark funds, and a pain point the first community-driven plans set out to fix.[12] Gradually, as utilities are undergrounded and roads are paved over, the scabby asphalt where people perished in their cars will be less visible. Touring the area with lawmakers after the fire, Town staff called for a moment of silence on a dead end plagued with fire fatalities. What may be harder and harder to see over time will still be felt by those who know.

I moved to Paradise from a town in the Bay Area called Moraga where I'd grown up and was then renting. The median home price in Moraga as of October 2025 was $1.575M, totally unaffordable for me then as now.[13] Moraga is a charming town nestled behind the Oakland hills where fog creeps in and sits low before rising and disappearing into mist. The nearest shopping hub is in Walnut Creek which was becoming gentrified with high-end condominiums and exorbitantly-priced clothing from brands like Prada and Gucci when I left. Back then, I loved window shopping and people watching in Walnut Creek where there were still a few classic stores from my childhood. I knew the best parking nooks and hidden restrooms where I could breastfeed Adela when she came along.

Walnut Creek has a sheen to it compared to Butte County, an impressive gloss with a hint of hard exterior perfection. The flowering trees, potted plants, and fountains for making wishes appeal to me. Downtown sits just outside of Broadway Plaza where city streets are lined with glittering trees, iconic restaurants, coffee houses, and frozen yogurt shops. It's walkable and somewhat affordable if you're snacking and looking for gently-used deals.

---

[10] Reported by the Butte County Sheriff during a tour of public safety facilities in Chico, CA, 2021.

[11] Reported by the Public Works Director, Town of Paradise, CA, 2024.

[12] Reported by the Public Works Director, Town of Paradise, CA, 2024.

[13] According to Zillow, October 12, 2025, https://www.zillow.com/home-values/32942/moraga-ca/?msockid=15de5b1c36ae688118264f7a37e26985.

The week before I moved to Paradise, I was pushing Adela's stroller across the street from Nordstrom in Walnut Creek when a driver waiting for the pedestrians to cross lost his temper. He was impatient and felt slighted by someone, maybe a pedestrian late to the pack holding up his car. I couldn't tell so I shielded Adela from the profanity and pushed on.

I still remember that altercation as an indicator of rage bubbling just beneath the shiny surface of Walnut Creek and the larger Bay Area. It was January of 2008 and quite possibly related to the mortgage crisis which hit the area hard. Either way, I reflected on that moment from Paradise weeks later where bank tellers regularly spent 20 minutes catching up with each customer about rummage sales and bridge games, while folks waited patiently in line for their turn.

Paradise was not a place to rush or be rushed. In fact, I only became aware of my tailgating habit when I found myself driving too closely behind a Paradise Chamber volunteer. It took me a few years to unwind my driving habits, but I can now enjoy poking along rural highways without a care in the world. When I hit I-80, my Bay Area skills kick in and I'm right back at it. My husband grew up in Chico and we agreed early on I would do the urban driving in our relationship.

I only lived in Paradise for three years but I met more people during that time than 30 years in the Bay Area. I went to Rotary meetings, joined the Paradise Community Foundation board, Chaired the redevelopment committee, organized Johnny Apple-seed Days with dozens of volunteers, shopped at Holiday Market, and became a regular in the McDonald's drive-thru where I ordered a large ½ sweet tea and ½ iced tea with extra ice. I tried the health food store, the Thai restaurant, and I loved the chips and salsa at Casa Paradiso.

When I lived in Paradise, I was a young woman with my whole life ahead of me. I was fearless in the safety of that small town tucked out of sight, and gave myself permission to experiment with dangly earrings and bangle bracelets without the pressures of high fashion in the Bay Area. I was a sparkle a minute with my sparkly child, and even though our lives were complicated by the custody commute, it was the simplest time, too. The strength I gained from my cheering squad in Paradise and my first non-profit executive job carried me into a satisfying career in Chico which now, in retrospect, marks my most confident career chapter to date. It took less noise, less social pressure, and less ladder climbing for me to find my happy place.

After the Camp Fire, I thought returning to the Town as Disaster Recovery Director might have the same effect but in reverse. Whereas Paradise once put me back together, I came to help put Paradise back together. But without the Paradise I'd known—those blissful friendships, the safety of the trees, my carefree spirit—there was no well to draw from. Paradise needed more than I could give.

Reference

Ewell PL (1995) The Oakland-Berkeley Hills fire of 1991. In: Weise DR, Martin RE (technical coordinators) The Biswell symposium: fire issues and solutions in urban interface and wildland ecosystems, 15–17 Feb 1994, Walnut Creek, CA. Gen. Tech. Rep. PSW-GTR-158. Pacific Southwest Research Station, Forest Service, U.S. Department of Agriculture, Albany, CA, pp 7–10. https://research.fs.usda.gov/treesearch/27400

Chapter 4
Work

A few years after I moved out of Paradise, I ran into the Town Manager in the ladies' room at a casino in Oroville during an economic forecast conference. Still well before the Camp Fire she said she thought it was a shame I'd left the Paradise Chamber. "You were our great hope," she said. I remember feeling unsure how to reply since the Town had defunded the Chamber and put the non-profit on rocky financial footing. I weakly excused myself and moved on, but I felt seen and grateful for her compliment.

In January of 2020, she called and asked me to come work for her as Disaster Recovery Director at the Town. I was quick to say no. I was finally making a decent living in an exciting state association in Sacramento, picturing myself climbing higher into larger geographical significance. I still had big dreams and had left my career in Chico without once looking back. Quite the opposite, I was lobbying my family to move to Sacramento to relocate our lives and shorten my commute. But her call stayed with me, as Paradise had all those years. I hadn't been back to the town since the Camp Fire, and hadn't begun to process my heartbreak over the devastation. Going back, let alone working there, was not an option.

Gradually, over the course of my long commute, I awoke to the possibility of helping the town that helped me so much. I agreed to meet with the Town Manager and mustered up the courage to drive up the hill for the first time in over a year. I'd been in the canyon to my parents' house which was a painful enough drive through charcoal lots and dying trees, but I hadn't seen the town.

On the morning of our meeting, Skyway crested into Paradise just as it always had. I passed Town Hall on the left, still standing and newly painted a fresh shade of slate gray. Across the street where the Chamber had been, with the cabinets we'd painted and the fancy desk Melissa donated to my spot by the window, was a weedy patch of concrete and dirt.

I knew I'd be shocked but it was more than that—I was disoriented—like my memory card was empty and my operating system was searching for data. I couldn't

K. Simmons, *Three Fire Mountains*, https://doi.org/10.1007/978-3-032-17343-0_4

take in what I was seeing and felt what I can only describe now as a visual rejection. A hard mental no.

When I eventually found the courage to look for my little blue house, I passed the grated dirt lot twice before a sliver of box hedge gave it away. I could see straight from the dirt road past where my house had been and the house behind it, all the way to the main road I'd never known was so close, hidden as it was by those trees.

A sense of home is dependent upon what's been added to the environment to create familiarity, comfort, and safety. When those additions are gone, so is that sense of place. Though I could see the patch of earth we'd lived on, without seeing what we'd lived in and around—the house, the trees, the driveway, the drying rack, the porch—my memories could not connect with the place. The little blue house lives on only in my mind.

For the pre-interview, I met the Town Manager in Starbucks which stood like an oasis on Skyway across the street from the Building Resiliency Center, a post-fire hub for contractors and owner-builders. Starbucks was shiny and new and filled with friendly baristas in familiar green aprons. I would come to witness a sensitivity with this Starbucks as it closed at the slightest hint of fire. When the air quality reached toxic levels during the North Complex Fire and outside work was suspended, Starbucks shut its doors for weeks.[1]

But on this day, it was open and I sat with the Town Manager listening to her talk about recovery. "But…," I asked, "is disaster recovery as grueling on the mind and heart as it sounds?" I wish then I'd listened more closely to what my question revealed about my fears. She assured me it was the most positive, gratifying work I could do, and I believe that's what she thought because it's also what I've grown to think, interspersed with the frustration, rage, and grief.

I was honest about my doubts that I had the emotional wherewithal to face the destruction day in and day out, and she encouraged me to focus on the restoration. With urging from former colleagues to return to work in the area, and to take on a behemoth job they seemed to think I was capable of, I ignored my inner voice and accepted. I had accepted a job with this many doubts once before, just out of college, and lasted three weeks.

For me, disaster recovery is not a straightforward gift of good will. It is a battle. Federal, State, and local governments do not see eye to eye and walk arms linked toward a shared goal. Unfortunately, there's a reason for that. Disaster recovery is a messy check and balance system of funding exchanged between all levels of government, where one brings the appropriation and the regulations, another layers on policies and procedures, and the last implements projects in hopes the funding will come.

---

[1] Personal observation between 2020 and 2021 due to significant fire and smoke events.

4.1 Hard Things

I have a theory about why State disaster workers are reassigned every few months, shipped off to a new disaster to work in response and early recovery. For the most part, the skill sets needed for response and early recovery are trainable, repeatable, can scale, and most relevant within a limited time period. That doesn't mean they're not taxing, which is where relocation comes in, a fresh opportunity to make a big difference somewhere else before burnout sets in. Long-term recovery is a whole other beast no one arrives four years later to take over.

In November of 2023, I attended the Rural Voices Coalition of California (RVCC) annual conference at the Stanford Sierra Center near Tahoe. It was a hopeful scene: incredibly passionate professionals from rural-based land management agencies coming together to talk about prescribed burning, the Farm Bill where funding for forest management sits, and other federal and State policies and procedures that govern land stewardship. I thought these might be my people, and I was delighted to be there.

Before I arrived at the Center in Fallen Leaf Lake, I curved through fire-scarred forests along the Highway 50 corridor where the Caldor Fire burned over 220,000 acres across three counties in 2021.[2] The closer I got to my destination the taller and wider a smoke plume grew like an unwelcome sign. It was a brown plume, not black, and it dissipated pretty quickly above the horizon suggesting a controlled burn, but my mind was already running away with me as I drove deeper into the untouched forest. On the eve of a much-needed professional adventure, here were the panicky feelings of home.

As I suspected, the land stewards were conducting a controlled burn. Over the next few days, the smoke held relatively low to the ground and settled on the lovely, placid lake. But driving in on that one-lane road I estimate I passed at least three traffic-controlled stops due to utility pole replacement, crossed one single-lane bridge, and saw nothing but smoke in my rearview mirror. This is what we tell people not do to. Three strikes of evacuation risk and I'm out.

By the time I parked in the conference center, I was doubled-over in tears. I recovered enough to get out of my car and press the registration staff for information. They confirmed it was a controlled burn and apologized for the inconvenient traffic stops occurring simultaneously on the one-way-in-one-way-out road. With my sunglasses obscuring my teary eyes I gave them some heavily trauma-laden advice to email such conditions out to registrants in advance to avoid panic-stricken arrivals such as mine.

The fog of trauma never fully lifted during the conference because neither did the smoke, but I engaged in the ideas and case studies with registrants and speakers, sharing meals and breaks with attendees. I sat next to a speaker from Sonoma who spoke about their prescribed burn program following their successive fires, and her

---

[2] CAL FIRE Incident Report, Caldor Fire, last updated 2/26/2023, https://www.fire.ca.gov/incide nts/2021/8/14/caldor-fire/.

experience evacuating her kids twice. During a federal advocacy session, I talked about the need for more study on long-term recovery protocols and procedures.

Response is a well-oiled machine, early recovery follows relatively similar tenants, but there's no rubric for long-term recovery other than a few academic publications on various roadblocks and approaches. At the conference, I envisioned an Incident Management System for long-term recovery, beyond the Emergency Operations Center (EOC) and Disaster Recovery Operations Center (DROC), and wrote about it on a poster paper collecting our ideas. The speaker read it aloud to the session attendees as one of the most compelling concepts which was nice to hear; however, there were so few of us working in long-term recovery that the idea did not gain more traction.

My theory around the revolving workforce post-disaster comes down to avoiding burnout. One of the issues we talk about at the County is how burnout can lead to a loss of institutional knowledge. One could say that Butte County is prepared for the next big fire because we now have a skillset built off of the last. The thing is, losing people to retirement, burnout, relocation, you name it, we're no more prepared than we were the last time. New folks show up on the doorstep of disaster with energy and willingness to learn—just like I did—and the cycle of doing hard things continues for those who remain to repeat, iterate, repeat.

4.2 County of Butte

A neighbor who worked for the County used to invite me to go for walks around the block. A few times, even before I worked for the Town of Paradise, she asked if I would come work with her. I always said no—I had an exciting opportunity in Sacramento and then I was giving back in Paradise. But as we kept walking, she changed my mind.

On my second day of working for the County the Dixie Fire ignited. A few days later as the incident grew, we assembled in the Emergency Operations Center in Oroville with our overnight bags. I carpooled with my neighbor who showed me the ropes. Information Systems fired up the monitors and hooked up all the computers, County pros found their familiar seats at the Section desks, and we got to work.

I trained in the position of Liaison which maintains communications between the EOC and the Incident Command Post (ICP), and between the County and cooperating agencies like utilities, other jurisdictions, school districts, and non-profits. Over the next few weeks, I went to the Silver Dollar Fairgrounds in Chico which serves as the ICP for Butte County's "big ones."

CAL FIRE moved into the Fairgrounds with their trucks and trailers and set up their command center, every vehicle and temporary unit clearly labeled with its purpose and function. At each morning briefing, hundreds of firefighters sat through the fire weather report, safety report, operations report, comms report, and roll call. Days into the fire, the number of firefighters climbed into the thousands as trucks packed the site from across the United States.

During those briefings I overheard weary conversations, snippets of, "I can't believe we're here again."[3] One day, as the massive fire shot slow-motion thunderheads from the mountains behind us, a plume of black smoke rose from across the street. Over the sound of the Dixie Fire briefing, sirens advanced toward us from throughout the city. The CAL FIRE Incident Commander raised his voice to be heard over the fire engines roaring nearby.

Are we burning down, too, I wondered in weary apathy? Nearly, was the answer, as an RV blasted and cooked across the street, a surreal smoky sight given the hundreds of thousands of acres burning behind us.

4.3 The Fellowship

When I first started the fellowship, I was under the impression that Butte County is recovering too slowly from the Camp and North Complex Fires. As of the 5th anniversary of the Camp Fire in November of 2023, 14% of the homes lost in the unincorporated area had been rebuilt.[4] In the North Complex burn scar, 4% had been rebuilt after three years.[5] I walked into the first cohort gathering on the Stanford University campus feeling like a failure at recovery.

What piqued this investigation was seeing emails from Stanford Impact Labs (SIL) that I deleted at first thinking they were spam. A week before the application deadline I took a closer look and saw a very inviting prompt asking what issues we're trying to solve at work, and how we might use data to solve them. County staff was in the midst of preparing an extension of Chapter 54 for the Board of Supervisors to consider, the urgency ordinance for the North Complex Fire. Chapter 54 allows survivors to remain on their properties in temporary housing without an active building permit under certain conditions. Standard code requires an active building permit for a property owner to live in an RV on their property during construction.

I forwarded the email from Stanford to Meegan Jessee, our Assistant Chief Administrative Officer, and asked if she'd take a look. I explained a quick outline of what I was thinking for the prompt. She had a few questions, we went back and forth, and she supported the idea of applying. I remember bringing her the letter of support to sign and, feeling a little giddy, said, "thanks for doing this, I love a good Hail Mary."

I was selected as a finalist, interviewed by a panel of experts and SIL staff, and awarded the fellowship. Before long I found myself nervously navigating the Stanford University campus looking for our meeting room. The breakfast outside tipped me

---

[3] Overheard numerous times from first responders during briefings at the Dixie Fire Incident Command Post.

[4] Camp Fire rebuild statistics provided for the fellowship by Dan Breedon, Planning Manager, Butte County, 2023.

[5] North Complex Fire rebuild statistics provided for the fellowship by Dan Breedon, Planning Manager, Butte County, 2023.

off and I walked in during self-introductions, delayed by not one but two stalled vehicles in center lanes going through San Francisco.

Cohort Gathering, November 30–December 1, 2023

Our first assignment that day was to draw a vision postcard of what our community would look like if our projects were successful. This was easy for me. Ideally, a resilient community is, as I said, the period at the end of every sentence that starts with disaster. I picked up the crayons and pens and drew a lovely idyllic scene with underground utilities, homes spaced apart with plenty of defensible space, interconnecting evacuation routes, plenty of water providing ample recharge, even a site for carbon injection though I only understand the idea conceptually. Dotted throughout my landscape were goats and cows grazing—nature's finest weed control.

When asked to share our problem statements I explained the destructive scale of our fires and stated we are recovering too slowly. I rattled off the stats and drew a chart on the whiteboard showing the progression of the rebuild over time. I talked about the layering of disasters—spillway incident, fire, pandemic, drought, more fire—and the compounding effects. My cohort fellows had equally pressing topics dealing with similar themes: vulnerable populations, systemic disadvantages, equity, programming, services, grants. I listened with genuine interest to their stories and problem statements, and by the end of the two days I was as hooked on their endeavors as I was on mine.

Late in the first day, as we were talking through how to form up our research, a lightning bolt struck my brain, the first of many. I'll call this the tide of obviousness that would roll in throughout my research.

I was sharing the names of the federal and State agencies we work with, and the role each plays in recovery. I explained the Federal Emergency Management Agency (FEMA), California Governor's Office of Emergency Services (Cal OES), the California Department of Resources Recycling and Recovery (CalRecycle), the U.S. Department of Housing and Urban Development (HUD), and the California Department of Housing and Community Development (HCD). I described their insistence on local control and how this does not appear to play out in my perspective, with the highest-stakes decisions made the highest up. I noticed a feeling of agitation in my body as I outlined these contradictions, a feeling of shrinking and suffocating at the same time. I'd been telling myself we should be so much further along, but the agencies making funding decisions about our recovery were not adapting to the barriers we could see from the local level. Why was my recovery meter set on "too slow" when no one—not even huge agencies responding to disasters all over the country—could tell us what the pace of recovery should be…or how to do it…as they make decisions and call it local control. Disasters are local, recovery is not.

My second lightning bolt: what in the world does recovery even mean?

I realized if I couldn't answer that question, was it fair to say we are recovering too slowly? Had FEMA sat us down and explained that by year five we'd be "here" on a linear route to recovery, with this many homes rebuilt and this much infrastructure restored? No. That's not even plausible when many of our largest infrastructure and staffing projects are held up in FEMA review. Had Cal OES provided us with metrics to evaluate and advance our efforts so we could keep pace with expectations? No, not when the repayment process for our very first early recovery effort, debris removal, was still unresolved five years later. In disaster recovery, nobody knows what they're doing with regard to the big picture, because there isn't one.

And…expectations. Whose expectations are the agencies working in recovery trying to meet? Survivors have expectations, Board members have expectations, the funding agencies have expectations. Are any of them the same? Have expectations ever been articulated side-by-side to reveal their differences, except during public meetings when fire survivors ask for what they need? To a person, recovery means different things.

I walked into the fellowship with a firm belief we are recovering too slowly from disaster, and walked away with my mind split apart. Who sets the pace for recovery, and who gets to decide what recovery is?

4.4 Defining Recovery

After the cohort gathering, I sent out a series of interview requests to those working in and around recovery. I asked local government staff, local agencies and non-profits, then I went up the chain and asked for time with statewide associations advocating for housing policy. I even sent a few invitations to organizations well outside of California but those went unanswered. In addition to connecting with people working in recovery, I went digging for a definition of recovery. I'd certainly been in and out of these federal and State documents over the years—FEMA's National Disaster Recovery Framework, HCD's policies and procedures, articles, guidance—but this time I was on a very specific mission: who holds the definitive definition of recovery?

Turns out no one. The most common definition of recovery describes the act of recovering, not what recovery is. The federal and State definitions, which are different, talk about capacity and coordination. According to FEMA: "Recovery is a cyclical, interdependent process where response, rebuilding, and mitigation often overlap. Progress may be disrupted or reset by new disasters, requiring flexible, adaptive approaches."[6] This definition tells me how to *think* about recovery, but not how to do it.

When I asked Professor William Siembieda, PhD and Professor at Cal Poly San Luis Obispo, to define recovery, he sent me a few references from disaster recovery publications:

> In a sentence, we believe that recovery of a city system means becoming a viable, adaptable system with a new normality in the post-event context. The establishment of viability in the present and for the future is the critical variable that defines recovery. It means that the system has a developmental capacity projected to result in continued self-sufficiency and

---

[6] FEMA National Disaster Recovery Framework, Third Edition—Amended, December 10, 2024, page 2.

that its key institutions are coping with and adapting to changing aggregate needs of the functions.[7]

And:

Recovery is a complex, multidimensional, nonlinear process. It involves more than rebuilding structures and infrastructure; rather, it is about people's lives and livelihoods. The process has no clear end point and there is not necessarily a return to what existed before.[8]

When I pressed Professor Siembieda on the inherent challenge of recovering without a universal definition he said, "Quantification is NOT common in definitions of recovery. This is because most practitioners find it is difficult to establish a meaningful measure against a time line. If you find a useful 'quantitative' definition of recovery, let me know."[9]

I understand what he means: as soon as you say what recovery is, you're wrong.

I googled 'recovery' and found: (1) a return to a normal state of health, mind, or strength, and (2) the action or process of regaining possession or control of something stolen or lost.[10]

Is recovery a return to what used to be or what has to be? After wildfire, these definitions seem to be at odds.

My dearest career mentor once told me a goal is not a goal if the outcome can't be measured. She told me I was wasting my time with ideas and visions if over time my progress couldn't be measured. Do I tell my children to learn as much as they can in school or to get good grades? Well, the former in my case, but how do we know children are learning well if teachers can't measure outcomes without the grading system?

I think we can agree on a few recovery basics. A non-linear process, sure. Complex, multi-dimensional, absolutely. I would throw in words like non-sensical, counterproductive, and heartbreaking. Quoting one of our longest-standing CDBG-DR disaster grant reps from HCD, disaster recovery is, "pushing a pig through a python."[11] But how long should that pig pass through a python, and what happens if it gets stuck?

I became curious about the definition agencies use to measure their own recovery outcomes. I interviewed a former Executive Director of the Butte County Housing Authority, and he said recovery is "disaster patronage."[12] Meaning, recovery is throwing a few well-intended dollars at a gaping hole of need and expecting people to be grateful. In a combined interview with Cal OES and HCD, staff described recovery as "a down payment," meant to restore the tax base or stabilize the community enough to make strategic decisions. In both definitions, recovery is a start but certainly not a complete restoration. Different definitions, similar results.

---

[7] Alesch and Siembieda (2012).

[8] Johnson and Hayashi (2012).

[9] Email correspondence with William Siembieda, Professor of Planning, CalPoly, 2024.

[10] Google's English dictionary provided by Oxford languages from Oxford University Press.

[11] Heard numerous times in standing meetings with disaster reps for HCD.

[12] Fellowship interview, Former Executive Director, Butte County Housing Authority, 2024.

If recovery is a tiny first step, what do we call the steps that come after? And once we take the step we call recovery, do we consider ourselves recovered? If the total spend-down of all federal funds is supposed to represent a down payment, and we're left with a restoration of, say, 15% of our homes at that point, do we consider recovery complete and the community recovered? So many questions.

For large scale disasters that qualify for federal funding, it appears we use the word recovery to loosely describe the period in which those funds are spent. How do we then define recovery following smaller scale disasters that don't meet the federal assistance threshold?

Federal disaster recovery funds are highly restricted to be used on housing, infrastructure, plans, services, and mitigation. And those funds, in our case post-Camp Fire, come years after the disaster occurs. Is the pass-through of those funds to the projects that hit the ground five to ten years after disaster our only act of recovery? And what if those funds don't actually lead to the individual recovery of survivors, as is often the case?

In an interview with the Chico Enterprise Record newspaper in 2023, I was asked about recovery. The reporter asked whose responsibility it is to make everyone and everything whole after fire. I explained there are different kinds of recovery that fall into generalized buckets. One is 'individual recovery' which occurs at the household level and is led by a single person or family to utilize existing funds and capacity to rebuild or relocate. Individual recovery leads to a number of different outcomes. Not all recovered households look the same, and not all households recover.

Local governments may be eligible for federal funds to rebuild and restore and, in some cases, expand the physical infrastructure. This is a big driver of 'community recovery.' These projects include water and wastewater system repairs or upgrades, road repair and reconstruction, subsidized housing funding meant to match other sources, evacuation planning, radio receivers for emergency alerts, essentially large public projects meant to scale across the impacted areas and far beyond for mitigation purposes. These projects are not led by a single person nor even a single government. They result from braiding federal funds and law, State policies and procedures, and local policies and Board actions.

Community recovery is also the result of motivated individuals and groups who collectively restore beloved community art pieces, spaces, and landmarks which are often ineligible for federal recovery funds. These are artistic, cultural and historic treasures the federal government doesn't see as necessary for the restoration of the tax base, and/or are not publicly owned. Many people in impacted communities believe these projects are necessary for healing and wholeness.

Looking back on this interview, I see that I organically knew then what I've rediscovered in this fellowship: there is no single definition of recovery. It depends on who you ask, what you're asking about, and who is responsible for it. During the fellowship I laughed over email with my cohort mentors about the fact that I now refer to myself as "Captain Obvious," discovering things I've been responsible for teaching others over the years.

I find it so interesting that if we already know recovery can't be defined or measured beyond the weight of a million different recovery-like acts, why do we hold ourselves and our communities to certain standards and expectations? I think the simple answer is that we see unbearable prolonged human suffering and no one wants to face the truth that recovery won't happen for every person and every place.

When a Cal OES staff member said "down payment" during our interview, no one flinched or rebutted. If it is generally accepted that recovery is a down payment funded by the federal government, then it must be true that the community is expected to come up with the remainder well beyond the local match. If it isn't immediately known post-disaster if the community can carry the mortgage, even nominal payments amortized over decades, then I wonder if the down payment is a bit of a test. Watching what has happened with the defunding of the Camp Fire Owner Occupied Reconstruction program, I'd say it is.

The funds we manage at work, CDBG-DR, are meant as "last in" funds to cover unmet needs. They are congressionally appropriated shortly after FEMA Major Disaster Declarations and are administered by the U.S. Department of Housing and Urban Development (HUD).[13] In California, the State Department of Housing and Community Development (HCD) is the recipient of these funds and here's my take on what they do next:

- They gather data to understand the uses and allocations of the funds. i.e., determine how much money should go to each category of need, and the ways in which the funds should be spent by the State or by the jurisdictions impacted by disaster;
- They look at the number of impacted homeowners and renters, and the rate of insurance to determine if there is a gap between available resources and the current cost of replacement to rebuild single-family homes;
- They look at the number of businesses impacted, their insurance claims, and the number of recovery loans administered;
- They look for data to validate how much money is needed when all other funds are expended.

Ultimately, HCD is constructing a defensible, transparent, data-driven explanation that yields a dollar value of unmet needs in a variety of categories that will be partially covered by CDBG-DR. Action Plans are the State's strategic policy document for the implementation of CDBG-DR appropriations, and must be approved by HUD. These plans take years to develop and go through several rounds of public comment before approval. Over the ensuing years of recovery, Action Plans can be amended a number of times, both substantially which requires another public comment period, or insubstantially which does not.[14] One way a local government can express frustration productively is to utilize these public comment periods to document the challenges of utilizing the funds as intended.

In the case of the Camp Fire, CDBG-DR represents an incredible sum of money exceeding $1 billion dollars meant to aid all eligible jurisdictions impacted by the

---

[13] Action Plans and Federal Register Notices (FRNs), CA Department of Housing and Community Development, https://www.hcd.ca.gov/funding/dr/action-plans-federal-register-notices.

[14] Action Plans and Federal Register Notices (FRNs), CA Department of Housing and Community Development, https://www.hcd.ca.gov/funding/dr/action-plans-federal-register-notices.

DR-4382 and DR-4407 presidentially-declared disasters of 2018.[15] The Camp Fire is included with the Woolsey Fire in Malibu, and both Butte County and Malibu among other jurisdictions are eligible for CDBG-DR funds that are highly restricted to meet unmet needs identified by the State.

I will go into recovery planning later but here's where the federal and State levels diverge from the local government level, making a cohesive recovery somewhat challenging to achieve. Federal funds, which are the largest single source of funding for recovery, are highly restrictive and require a great deal of capacity and training to administer correctly, and ultimately may not match with a local jurisdiction's needs. If a community envisions its future in a recovery plan and federal funds do not support that plan, or at the very least are difficult to shoehorn into that vision, a community can be left with diverging recovery processes no single agency has the power to bridge. The Town of Paradise has a community-scale sewer system in its recovery plan that may be impossible to fully fund with available resources. This is why adaptation and flexibility are so critical during a long recovery process, which can be a hard pill to swallow for a community that needs very specific, costly infrastructure to realize its vision.

I imagine when people look at ghostly images of wildfire destruction, they think first of humanitarian aid which is necessary for meeting immediate and individual post-disaster needs like shelter, food, clothing, medication, and other basic supplies. Well-intended donations made by individuals and businesses (and some celebrities) flow in to impacted communities meant to directly support fire victims. After tragedy, people want to help people.

Humanitarian funds given charitably by donors are among the most flexible. They are often overseen by local non-profits or foundations governed by a Board of community members who set criteria covering a wide net of possibilities. The non-profit might come up with the ways in which the funds are meant to be spent, and/or those who apply might have the ability to define their needs. Either way, once those funds are awarded, they are likely not tracked for the next 55 years and doled out in dollar-for-dollar reimbursements through monthly financial and activity reports that justify every expense, as is the case with federal funding. Local organizations stewarding charitable recovery funding can answer the call of human need without measuring that need beyond what they can see given their proximity to the impacted area.

Local governments using federal recovery funds, on the other hand, have data to understand the local impacts and little control over how to spend the funding short of prioritizing eligible projects.

---

[15] State of California Proposed Action Plan for Disaster Recovery from 2018 Disasters Master Action Plan, November 2024, https://www.hcd.ca.gov/sites/default/files/docs/planning-and-com munity/18dr-master-action-plan-original-to-apa-7.pdf.

4.5 Down Payment

Disaster recovery funds coming from the federal government in the form of FEMA Public Assistance or CDBG-DR are public dollars. These funds are incredibly necessary for recovery, but they are not governed by local non-profits made up of Board members living in the community and witnessing the need first-hand. Federal recovery funds are governed by FEMA or HUD, headquartered on the opposite coast of our burn scars. Despite federal and State representatives who arrive when the fire occurs, funding decisions are not made at tables where those representatives meet with locals. Conversely, local governments must plead their cases in one public comment letter after another which, to be honest, feels like shouting into a void.

I imagine if the federal government is making a down payment, they are going to first run the community through a credit check to see if there's a robust enough economy to support those mortgage payments once the deposit is down. To us, the inability to qualify or use those funds locally feels like recovery tragedy layered on disaster tragedy, whereas it's likely stemming from policy decisions made by strangers.

Take owner-occupied housing reconstruction loans (OOR) administered by the State according to the rules they set in the 2018 and 2020–2021 Action Plans for the Camp Fire and North Complex Fire, respectively.[16] Typically, the application period opens a few years after the disaster when insurance proceeds, if there are any, have likely been spent on emergency housing like RVs. Households need these OOR funds because they do not have enough resources to rebuild on their own. If they've already spent their insurance proceeds—which are considered funds for reconstruction—on RVs or temporary housing, they have to come up with another source of funding to essentially "reimburse" what they spent on that RV in order to qualify for the OOR loan. This is called duplication of benefits.[17] Funds from the federal or State government must not duplicate other funds meant for the same thing. Strike one against the homeowner who, out of desperation waiting three years for a rebuild loan from the government, spent their insurance proceeds on temporary housing.

Let's say in the Action Plan the State carves out $175M of $1B in CDBG-DR funds for OOR, and after three years qualifies 11 households of the hundreds still in need of housing. Those 11 households will work with the State through the rebuild process and will eventually move back into safe and secure permanent housing.

The rest? Households that cannot be helped by the funding of last resort, say because of duplication of benefits, are then left to their own resources to recover. As we know too well in Butte County, any delay in permanent housing prolongs displacement and increases the risk of disaster-caused homelessness for survivors. In the case of post-wildfire remote rural living, homelessness is nearly invisible unless you know where to look.

---

[16] Action Plans and Federal Register Notices (FRNs), CA Department of Housing and Community Development, https://www.hcd.ca.gov/funding/dr/action-plans-federal-register-notices.

[17] Duplication of Benefits as defined in the Stafford Act, September 24, 2024, Section 312, pages 28–29.

4.6 Windfall of No

It's important for me to express that while I have experienced the challenges of administering federal funds after disaster, I am keenly aware that recovery in any form is even more challenging without them. We have seen our own communities flounder in recovery after fires without a federal declaration, such as the Park and Thompson Fires in 2024. Butte County relies heavily upon FEMA PA and CDBG-DR for federally declared disasters, and feels the deep cut of recovery when smaller scale disasters do not trigger federal assistance. Every effort to articulate the challenges of utilizing these funds is a sincere attempt to identify the very specific policies and procedures that might be adjusted for better, more efficient use. Everything I say in this book related to funding challenges, I have also said directly to the agencies I work with in recovery. Observing my own aha moments during the fellowship, I have come to appreciate how helpful it is to explain the barriers we find nearly impossible to see with our own eyes. As with the many contradictions we contend with in recovery, gratitude for federal assistance and the frustration in using it are cornerstones of our efforts.

I hadn't even left the Town of Paradise yet when I started calling the allocations made for disaster recovery a "windfall of no." We'd hear about gigantic sums of money made available for survivors and local government, often reported in the media as a "windfall" and as we'd start to read the fine print, maybe paragraph three out of 400 pages, we'd discover those funds may never touch down.

Duplication of benefits is a big one. So is income qualification. Low to Moderate Income (LMI) data is often pulled from pre-fire census numbers which were based on the presence of employed people with insurable assets—a totally different town than what was left after the Camp Fire.[18] Where are those people, those jobs, those homes after fire? Not where they used to be.

Paradise lost one of the largest employers in the county paying some of the highest wages for the highest skilled work. Feather River Hospital issued 1,300 layoffs after the Camp Fire and was ultimately unable to reopen.[19] Doctors, nurses, medical personnel, administration, facilities, each and every person had to find another job whether they were relocated by the company or not. This took an immediate but invisible toll on the recovery as those relocating professionals sold their lots or standing homes and left for jobs outside of the area. The exorbitant need for housing after the fire also motivated early retirements and out-of-state relocations for many

---

[18] Qualification requirement for several sources of federal funds following the Camp Fire, such as Community Development Block Grant Disaster Recovery funding administered by the CA Department of Housing and Community Development.

[19] Urseny, L., Adventist Health Finalizes Layoffs at Feather River Hospital, February 13, 2019, Chico Enterprise Record, https://www.chicoer.com/2019/02/13/adventist-health-finalizes-layoffs-at-feather-river-hospital/.

professionals who saw an opportunity to sell their houses at demand-driven peak prices.[20]

In California, HCD acts as a pass-through for federal funds from HUD. When I was at the Town, HCD leadership agreed to sit on a zoom call with us while we made the case that post-fire data should be used for determining LMI eligibility. The State agency director watched while we pulled up drone footage scouring the barren, broken, ashen landscape where street after street had maybe one or two standing homes, if any. It was equivalent to the "earth" view of the Damage Inspection Report map.

"Doesn't this area qualify for recovery funding meant for vulnerable, disadvantaged populations?" we asked incredulously. The answer was no.[21] Due to the timing of the last census, income limits were based upon a population that was no longer there and unlikely to return anytime soon, which seemed as ridiculous to me as administering treatment for a diagnosis found during autopsy. Whether or not it was intentional, the barrier felt like a good reason to deny assistance and walk away.

4.7   Recovery Rage

Back to the down payment, I expect the stewardship of taxpayer money at the federal level to mean those funds are not poured into bad investments. But there we set up the catch-22 of recovery which is a sound but complex investment. If the federal government or State observes in testing the down payment that there's not a penny to be put toward the mortgage for miles, will that windfall be made available to the community, or will the barriers continue to strategically appear? As a taxpayer, I hope no funds are used with wild abandon. As a human being sitting in the midst of suffering, it is unbearable to think of even the slightest withholding.

I arrived in the field of disaster recovery a year and a few months after the Camp Fire, so it's quite possible that when recovery conversations first occurred these hard truths were spoken with local officials. I picture a long, shiny boardroom table where the federal government sits at the head and explains they will give a little and see if the community can cover the remaining 80–90% of the sale price. Whether nor not this conversation occurred, and I doubt it did in the way I am imaging which looks a bit like a Far Side cartoon, I have picked up these observations over the course of recovery.

The magnitude of my personal disillusionment says a lot about the federal and State governments, but more about the level of earnestness and trust I once had for human beings doing the right thing. I remember asking the new Town Manager in Paradise, who arrived shortly after I did, if I was supposed to be in a rage all the

---

[20] Observed from many friends and family following the Camp Fire, confirmed by realtor interviews conducted during the fellowship.

[21] Observed during a meeting between the Town of Paradise and the CA Department of Housing and Community Development following the Camp Fire.

time. He said absolutely. If I wasn't enraged by the mind-numbing contradictions of recovery, and the unfeeling pushback from outside government agencies, I wasn't paying attention.[22]

A few months after the Camp Fire, my parents and I met in a local coffee shop to comb through their insurance claim. I was working remotely for my job in Sacramento and they were popping in to my "office." Minutes into the conversation my mom was in tears about how little her insurance adjuster cared about her situation. She wanted to send him a letter admonishing his apathy and encouraging him to care about her loss. "No, mom," I said, "don't waste the emotion that letter will take you to write, it isn't his job to care. His only job is to help you recoup as much as you can from your losses." Following the LA firestorm in 2025, California Insurance Commissioner Ricardo Lara appeared to be somewhat correcting this lengthy itemization process by asking insurers to pay out approximately 30% of every claim that demands this step.[23] This would have helped my parents immensely back in 2018–2019, but the realities of slow insurance payouts were not as well known.

Imagine if my mom's insurance agent felt real pain for her and the dozens, maybe hundreds, of wildfire survivors weeping on the other end of his phone line. He could and probably did show caring and compassion, but I've found the effects of secondary trauma—the act of taking on someone else's heartbreak as your own—can linger for years. Whatever boundaries he kept, whether by training or by choice, likely shielded his heart from the repetitive strain of the job. For my mom to put this much work and time into her insurance claim, to withstand the painstaking act of remembering all she'd lost, she needed her adjuster to appreciate the meaning in the list not just the value.

I wasn't strong enough to show deep compassion for their losses at that time either. My dad died in October of 2016 and I closed out his estate in October of 2018, just one month before the Camp Fire. I spent two years selling, moving, bartering, tossing, donating, and storing his house full of clothes, books, furniture, tools, boats, cars, motorcycles, you name it—everything left when a life in full swing is interrupted. I couldn't bring myself to care about the stuff my mom lost in the fire, it was grief on top of grief which is what leftovers feel like when there is a death in the family. I'd lost my dad and was left with all of his stuff; my mom and stepdad were very much alive and their stuff was gone. I preferred the latter and stuck to that lane. Either way, I lacked the language of trauma which I'm only beginning to understand now that I need it.

I have never been prone to rage and was entirely unfamiliar with it when the fire happened. Sitting there with my mom and her list in what was supposed to be a soothing place of bakery scents and warm drinks I could feel nothing but rage. I tried to find my value by producing an orderly list they could follow, writing down steps that must be taken, making recommendations for who to call, talking them through the fact that the smoke remediation company might not have returned their call yet

---

[22] Observed during a conversation with the Town Manager for the Town of Paradise in 2021.

[23] Presented on a webinar hosted by the CA Department of Insurance after the LA wildfires in January of 2025.

because they were overrun. I did my best to be a stable, reliable presence for my mom and stepdad but that was it.

At that point they were still displaced, living with my sister in the house she rented from them in Chico. Not long after the fire, during this period of displacement, my stepdad was hospitalized with sepsis. He was a survivor helping survivors, trying to keep up with his real estate business and clients in the Paradise market. In the midst of the post-fire turmoil, he came down with a routine illness that quickly turned into something worse and he collapsed in the street. At the hospital, he sat hooked up to hospital monitors and IVs while pounding furiously on his laptop and shouting into his phone. He was trying to reconcile some of their finances and move money between real estate assets while having his life saved which, to me, seemed like an afterthought.

Looking back on this now I understand his adrenaline-fueled panic. He was in the business of housing and, overnight, the county lost more than 15,000 housing units to fire, 14% of its total housing stock.[24] Ken sold properties in Paradise weeks before they were destroyed, leaving his clients homeless while they were still moving in.[25] He and my mom own a number of single-family rentals in Chico that were now in desperately high demand. They had hard conversations about the ethics of releasing tenants from their leases so friends could move in to their rentals, or so they could move in themselves. Displaced renters in a housing market flooded with buyers and renters created its own wave of upheaval.[26] Ultimately, my parents decided not to break any of their leases and thankfully Ken made a full recovery.

I interviewed a residential realtor from Coldwell Banker who retired after experiencing the same chaos I saw in my family. In retirement, he told me he was looking forward to gardening and traveling and moving past the intense stress of real estate during the aftermath of disaster.[27] He and Ken retired at nearly the same time, five years post-Camp Fire, which marked a turning point for many professionals working in or adjacent to recovery.

Reflecting on my own state of mind when Ken was hospitalized, I was in no shape to help with disaster recovery on a family level, let alone at a community scale. I was a stressed-out mom commuting 70 miles each way to and from work, dealing with the fallout of the fire on my immediate family. My youngest sister was pregnant with her first child and housing our parents and my kids on and off at her house in Sacramento. My other sister had a toddler and was housing our parents on and off in Chico. I was overwhelmed and paying attention only to what needed to be done in the moment, not what needed to be felt and processed related to the Camp Fire for the long term. There was no time for that, nor did I even know where to begin.

---

[24] Camp Fire Regional Economic Impact Analysis, Final Report, January 2021, Economic & Planning Systems, Inc., and Industrial Economics, Incorporated, page 98.

[25] Reported by Ken Dickson, Realtor for Keller Williams, in 2019.

[26] Observed from listening to several realtors after the Camp Fire, including brokers from Keller Williams and Coldwell Banker.

[27] Fellowship interview, Retired Realtor from Coldwell Banker, 2024.

From my home in Chico, I could see the FEMA local assistance center set up in a former Sears store across the street. RVs moved onto my neighbors' lawns and into nearby construction sites, sardine-like rows of travel trailers and 5th wheels on freshly-graded gravel. I think more than anything I could feel—because I heard about it from friends and family and saw it on the news and on social media and in every nook and cranny of Chico—the infinite heartbreak that comes with a disaster of this size. The fire was out, but it was everywhere, and it was not going away.

4.8 Involuntary Sprouting

Rodney, Heidi, and I drove up the ridge before I started my job at the Town to take it all in together. It was Rodney's first time seeing the town after the Camp Fire, and I felt it was important to see as much as I could in the safety of my family. I remember Rodney vocalizing his positive outlook as we drove, focusing on what survived and what was visibly being reconstructed. I shared his perspective while I tamped down my heartbreak.

Our youngest daughter, Heidi, was around 7 years old and sat in the back of the car. We drove above town limits into the unincorporated community of Magalia and from a pull-out we could see across to Sawmill Peak where the fire had scorched the trees. Not a single thing was alive between us and the peak aside from weeds, and I was devastated by the sea of destruction.

"How beautiful," Heidi said quietly, marveling at the sight, shocking me completely. I could see nothing but desolation and miles and miles of destroyed habitat but she saw something else. It made me teary, proud, and worried. I was thankful she could see what was left and still find it magical, and I worried her eyes were already attuned to the effects of fire. Heidi is wired with goodness, however, and I do believe she saw something pretty in the scenery that day, even if it was just the vastness between her and the mountain.

The local realtor I spoke to for the fellowship confirmed my theory. He said new buyers find Paradise just lovely. They can see the valley from their hilltop homes, sunrises and sunsets, big sky vistas, and twinkly lights from rebuilt homes around them at night. They look across the canyons to the ridges on either side of Paradise at stunning views.[28] I can certainly see the appeal.

Before the fire, homes were nestled deep within forested pockets, enclosed, shrouded, private. The ground was shaded and carpeted with soft pine needles that absorbed every sound and smelled like heaven. That feeling of being small and safe living among the trees—that was Paradise. Forests put life into scale much like the ocean. Above Paradise in Magalia, in neighborhoods that survived, homes still have that summer camp cabin feeling.

---

[28] Fellowship interview, Retired Realtor from Coldwell Banker, 2024.

I remember the sound of a giant pine tree falling late one night from my little blue house in Paradise. It must have been 100 feet tall and all I heard was slow-motion cracking and breaking before the ground shuddered and thundered like an earthquake. Being from the Bay Area I thought it was an earthquake, and I rushed out of my house into the silence of an unconcerned neighborhood.

The older I get, and the more I learn about the land, the more I understand the correlation between elevation and trees. Specific tree species grow at certain elevations, so you might have an idea of how high up you are when you spot one. A friend seeks out forests that mimic the elevation of the home she lost in Paradise so she can be surrounded by the same mix of trees and shade and scents.

I have always loved trees and in retrospect I see I took them for granted when I lived in the Bay Area. I enjoyed the oaks, eucalyptus, and cypress without a single thought of when the limbs would drop from disease, or if they were at risk of fire. I simply enjoyed their beauty as unchanging fixtures on the landscape.

In the Bay Area, I was attuned to the coastal fog and the bay views from the freeway system. Like certain cultures can explain snow, I can explain fog. I didn't realize how thickly the bay smells like ocean until I moved out of the fog and into the northern grasslands where everything smells like soil. Now, my nose can spot the ocean from 30 miles away and valley orchards smell like home.

I learned a lot about trees from the arborists and foresters working on the hazardous tree removal program in Paradise. They showed us post-fire signs of tree life and death. An arborist can analyze a burned tree and determine the likelihood of its survival within the next 18 months. They place this likelihood on a number of factors including the size of its remaining canopy and the softness and scaling of its burned bark. Some tree species can survive with only a third of their canopy intact, some need much more.[29]

Amazingly, bark tells the story of the fire that burned the tree. Fire tends to burn hotter and longer when trees are close to structures. I remember one forester pressing his fingers into the bark and explaining how hot the fire burned and for how long, based upon the surrounding circumstances, and why the fire didn't entirely engulf the tree. Flame marks on a tree trunk can be read like tea leaves, and this forester could see the fire by reading the tree. He'd worked on tree removal following many fires, and I could see in his eyes and hear in his language a skillset I never knew existed. Trees tell the story of fire in a way nothing else can because when everything else is gone they stand and bear their scars.

A few years after the Camp Fire, hundreds of trees grew mossy and fuzzy along their trunks but had no leaves on their bare, blackened branches. I called these "toupee trees" for how they were trying to cover themselves but were missing leaves in all the right places.

"That's a dead tree," a forester said, pointing to a toupee tree and breaking my heart. "But…it has plenty of leaves sprouting from its trunk," I countered, "see, it's

---

[29] Observed from tours and field visits with arborists and foresters during Camp Fire Hazardous Tree Removal Operations, 2020–2021.

green!". "No, no," he explained, "that growth is the tree's last attempt at photosynthesis. It is involuntarily sprouting as it is dying."[30] Involuntarily sprouting as it is dying…nature's forced last breath. I explained this phenomenon when taking consultants and officials on tours of the town. They'd argue with me as I argued with the forester, wanting to believe the toupee trees were alive. There were so many of them, and they still stood so proudly, survivors as they would have the untrained eye believe.

4.9 The Obvious

I am sharing what I've learned the hard way which now seems quite obvious. Jennifer Gray Thompson from After the Fire told me, "Don't assume people know. Tell the story. Have the audacity to will it into being."[31] If there's one word I hear over and over about wildfire it is "understudied," so maybe this isn't obvious after all.

Disaster recovery has forced upon me the art of getting comfortable with the uncomfortable. The unknowing isn't hard, it's the compounding effect of uncovering layers and layers of the unknown that's mind-numbing. The limitless, infinite, inexplicable layers of uncertainty suffocate even the best of intentions. There is nothing repetitive, predictable, nor linear about disaster recovery. There is no mastery of a single process or program that reveals any more about recovery than what it says about itself. I was reading an academic paper on wildfire-driven migration recently and one page alone contained more than 30 research citations. What we know about recovery is pieced together from a myriad of sources.

Starting over mid-career feels like a setback given the steep learning curve. This has certainly been my experience leaving the cushion of non-profits for government recovery. I would guess most people who find themselves in disaster recovery do so without formal training. They may be compelled by disaster to jump into the field, or invited in like I was. Either way, entering the field requires a full reboot—a humbling, freefalling, start-from-scratch plummet into the unknown. In my experience, there are no hours you can take into disaster recovery from any other profession that will add up to the 10,000 you need on day one to be effective. There is only the first hour of the first day of the first week and so on.

At the Town and at the County I've been entrusted with supervising programs I've had to learn during supervision. Saying this out loud doesn't strike me as uncommon. In the working world, people constantly shift to oversee programs and services they've never performed themselves. There is something different about recovery, however, in the way the uncertainty never goes away and, for many staff, by the fact

---

[30] Spoken by a forester on a field visit during Camp Fire Hazardous Tree Removal Operations, 2020–2021.

[31] Fellowship interview, Jennifer Gray Thompson, After the Fire, 2024.

that emotions are heightened by their own disaster trauma. I'll never forget reconvening in our Emergency Operations Center (EOC) for a training after the Dixie Fire when someone in uniform shouted from the back, "I've already called my therapist!"

In disaster recovery, a feeling of responsibility for the public drives up the stakes. In my case, it's also our team's oversight of over \$245M in federal grants that far surpasses any non-profit I've ever led. My annual budget at the Paradise Chamber was \$168,000 back in 2008.

The phrase 'federal regulations' is intimidating on purpose, translated into layman terms to mean, "decisions are not yours to make, mistakes will cost you dearly, tread carefully." Complying with federal labor standards reminds me of the boardgame Operation. Touch a nerve and "ZAP!" Violating one requirement can result in a recapture of all granted funds, even after they've been spent. Policies and procedures differ from program to program and must be followed with demonstrable, reportable exaction. The Stafford Act is dense. Due diligence involves assigning authority during public meetings, debarment checks, site control, local policy drafts, and environmental review. Monitoring requires annual federal single audit reporting and decades of oversight.

The State is its own distinct layer, a pressurized space between local and federal governments, an entire atmosphere buffering rock from space. The State has its own personality, its own posture, its own rules in disaster recovery. It is designed to hold tension between federal obligations and local realities. We often refer to the State as a second national government given the size of California's economy, which means our small county is dwarfed by two world superpowers who are rarely on the same page.

Professor William Siembieda says taking federal money is like, "making a pact with the devil."[32] He refers to his research in Chile where the private sector swept in and rebuilt homes in less time than it's taking us to set up federal grant-funded programs for recovery. He says FEMA doesn't mean speed it means money, and there's no way we can have both. If we're taking in federal funding and expecting a fast recovery, we're signing up for disappointment.

I asked Tracy Rhine, a legislative analyst with the Rural County Representatives of California, if she believes the State is prioritizing housing construction in urban areas post-fire, to both de-prioritize rural living and to solve the homelessness crisis. The overarching decision, as she explains it from the State's perspective, is to funnel housing to urban areas where the pinch of homelessness is felt and seen more acutely than displacement caused by rural wildfires in areas where homes burn again and again.[33]

Over the course of my Stanford fellowship, Hana Passen, Director of Stanford Impact Labs and my fellowship mentor, connected me with an associate professor using data to improve community projects. Her work looks at how proposed public infrastructure projects inequitably disadvantage vulnerable populations, and how governments might use data to explore more equitable approaches so projects can be

---

[32] Written correspondence from William Siembieda, Professor of Planning, CalPoly, 2024.

[33] Fellowship interview, Tracy Rhine, Rural County Representatives of California, 2024.

designed differently, better, to reduce those impacts.[34] I appreciate these conversations because I find the academic approach to sweeping social issues fascinating. It is often guided by data and research we don't have access to in small local governments, nor the time to fully absorb. Working on the macro level also seems like a less emotionally taxing way to make a difference.

Sometimes, though, the delivery of well-intended research misses the mark. A group of researchers called us in Paradise a few years back to say they modeled out a series of rebuild scenarios to see which would reduce the risk of repetitive catastrophic fire the most. They concluded that if we rebuilt in the center of town, with densely packed houses on both sides of Skyway, avoiding building homes out to the ridges on either side of town, we'd be looking at the most sustainable, lowest-risk recovery.[35]

Sure, we said, a tickle of rage in the back of our throats, but the properties all the way to the ridge are privately owned. What do you suppose we say to those property owners about their right to build on private property without any exchange or incentive to move? And how do you propose we build dense housing in an area 100% reliant upon septic where lot size has to account for leach fields? And what about the local build-back policy the Town Council set that honors property owner rights to construct homes similar to what they lost?

On paper they had a point. If recovery was linear and sewer came before housing perhaps the density option was feasible. And consolidating housing along the largest evacuation route makes sense for the quickest exit, leaving the ridges most susceptible to fire less populated, which could benefit insurance ratings. On the cost and access side it also makes sense because centralized housing wouldn't require private road maintenance and locked gates. This is as straightforward as the master planned development solution.

But…this model leaves out key factors that drive reconstruction in a town like Paradise. It does not account for the preservation of personal choice after catastrophic loss. It assumes the wrong things about what government may or may not be able to do on behalf of their disaster-impacted constituents. And it completely overrides the capacity of the existing local infrastructure. For us, a little more than a year out from the fire, hearing that an impossible scenario was the only way to safely recover, left us dismissive at best and bereft at worst.

4.10 Destroyed Community

I shared with the associate professor a desire for institutions who are not in emergency management to recognize and adapt to a new definition of community: destroyed. This definition requires a recognition that communities not impacted by disaster are intact, which feels like a fleeting reality we should not take for granted anymore.

---

[34] Correspondence with Stanford faculty and staff during fellowship, 2024.

[35] Meeting with academic researchers while at the Town of Paradise, 2020–2021.

Federal and State funding eligibility is tied to a community's ability to leverage the proposed grant-funded investment to improve the quality of life for disadvantaged populations, also known as meeting a national objective. The objectives are prescriptive and generally require proof of servicing low to moderate income clients, households, or areas. After a fire, however, there is often not the presence of people of any income, nor households in areas they once were. Snapshots of the population taken only occasionally through the census fail to articulate or predict wildly changing demographics caused by disaster. Funding agencies rely on old census data even if it precedes disaster and doesn't tell the whole story.

I believe the field of disaster recovery would benefit from a destroyed community definition; one we can build a new suite of studies around to understand. Analysis of lingering Camp Fire Impacts by Chico State, shows that 89% of the total housing stock in Paradise was destroyed in the fire, and 96% of mobile homes and motor homes serving as housing.[36] This is an undeniable post-disaster reality that deserves a different kind of analysis for recovery. While I understand reaching into the trusty data toolbox of the census to measure pre- and post-fire demographics, it's as futile as holding up a colander to catch the rain. You can't measure what has fallen through the cracks by looking only at what's left. Where are the tools to predict the long-term effects of losing 89% of a community's housing stock to wildfire?

The millions of dollars available for recovery have to be used in such specific ways according to such specific standards that they're often not attainable for impacted jurisdictions. We'd be on a scoping call with hazard mitigation funding reps—think government shark tank—who were pleased with the volume of funds available to assist, only to shoot down our ideas one by one. As a new disaster recovery professional at the time, I pitched projects in the Long-Term Community Recovery Plan and other common-sense projects needed throughout the town. If I wasn't using the correct terminology, we'd have to pause for someone to read aloud from the federal regulations. I quickly learned disaster recovery is not the place for common sense, and was not ashamed to close my eyes in frustration on camera during the read-outs.[37]

Threading the needle of disaster recovery grant eligibility when communities are destroyed is not impossible, but it's certainly more complicated than the funding notices would have the public believe. And therein sets up public expectations that are reasonable based upon the amount of funding "available." In reality, local governments may not be able to use the funding on time or as prescribed, hence the desire to broaden non-disaster funding to address destroyed community needs.

---

[36] Jacquelyn Chase, Professor, Department of Geography and Planning; Peter Hansen, GIS Specialist, Department of Geography and Planning, Paradise Continues to Rebuild and Evolve Three Years After the Camp Fire, November 8, 2021, https://today.csuchico.edu/paradise-rebuild-and-evolve/.

[37] Observed on project scoping calls for Hazard Mitigation Grant Program funding administered by Cal OES, 2020–2021.

The unincorporated community above Paradise has a recovery plan.[38] It's not called a recovery plan but it represents the community's vision after Magalia lost 2,158 residential structures and a significant portion of its commercial district in the Camp Fire.[39] My belief is any plan developed within 10 years of a major disaster is a recovery plan given that it has to account for major shifts in population size, demographics, and reconfiguration of businesses and homes.

The Upper Ridge Community Plan focuses on what's possible in the commercial district of Magalia following a rezone for expanded uses like mixed-use residential. To achieve the vision, the area will either need ample septic capacity which takes a number of parcels out of commission to serve as leach fields, or a community sewer system.[40] We tried the community sewer system approach first since it opens the most doors for residential density and commercial viability.

In our scoping call with the State Water Board, they asked how many connections we had in the project area. Connections? The area is totally destroyed. So…none. But there's great potential for hundreds of connections that will prove most beneficial to the ridge if constructed to the capacity of sewer rather than to the limitations of septic. We're talking restaurants and multi-family housing which means the highest density housing would fall on the main evacuation route which makes the most sense for life safety. Isn't that what those researchers had proven out for Paradise? (Fig. 4.1).

The Water Board representatives declined the project because without existing connections the project was not eligible for funding.[41] Without connections we couldn't build sewer, but without sewer the ability to rebuild on septic would be significantly constrained which would decrease or suppress community viability. Catch-22.

From here we jumped on the familiar merry-go-round of advocacy. First, we asked the Water Board to adapt their funding criteria to our post-wildfire circumstances, and we offered adaptations to each criterion. We expressed our sincere disappointment in their decision and inflexibility, and took this experience on the road, laddering it all the way up to the Governor's office as an example of destroyed community disadvantage.

I didn't realize there was such a thing as an intact community until I worked in a destroyed one. Now that I know the difference, I see how it plays out in widening the equity divide between communities that experience disaster and communities that don't. I want everyone else to see the distinction too, particularly as the risk of catastrophic disaster increases.

If we had scoped the community sewer project for disaster recovery funding it might have worked, but because we went the traditional funding route it did not. I

---

[38] Butte County Upper Ridge Community Plan, produced by Butte County staff and consultants including PlaceWorks, adopted by the Butte County Board of Supervisors on March 8, 2022, https://www.buttecounty.net/460/Upper-Ridge-Community-Plan.

[39] Butte County Planning data, https://www.buttecounty.net/460/Upper-Ridge-Community-Plan.

[40] Upper Ridge Community Plan, Chapter 8—Utility Infrastructure, p. 8-3, https://www.buttecounty.net/DocumentCenter/View/2794/Upper-Ridge-Community-Plan---Final-PDF?bidId=.

[41] Project scoping attempt for the Small Community Wastewater Program funded by the Clean Water State Revolving Fund (CWSRF) from the State Water Resources Control Board, 2023.

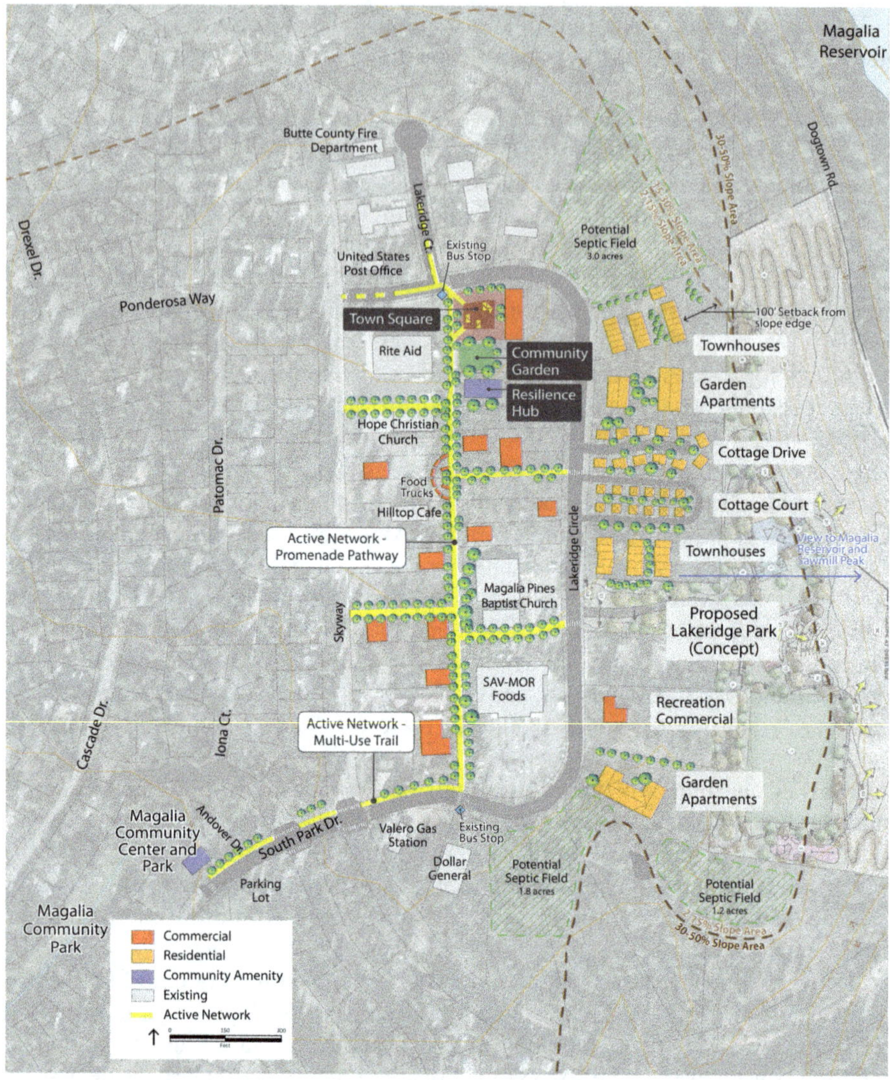

Fig. 4.1 Map of Magalia Concept included in the Upper Ridge Community Plan (4-3) approved by the Butte County Board of Supervisors on March 8, 2022. Produced by Butte County staff and consultants, PlaceWorks et al.

believe we need to unlock traditional funding for destroyed communities, particularly when disaster recovery funds are complex, have finite expenditure periods, and set up long-term recovery projects to compete for one-time funds. If traditional funds are off the table and recovery funds are limited, the pathway to funding recovery is prohibitively narrow.

4.11 Rural Life

It's no secret that homeowner's insurance in the State of California is becoming increasingly inaccessible and unaffordable by the day. As of this writing, big name companies are leaving the market, canceling renewals, or refusing to write new policies due to increasing wildfire losses and risk. United Policyholders reports that and unprecedented number of people are now on the Fair Plan, the increase at this point likely over 300%.[42] Insurance affordability and availability is an inconvenience in urban areas where homeowners have fewer options, but a crisis in rural areas where homeowners have none.

Currently, and more than once, AT&T has filed with the California Public Utilities Commission for relief from its obligation as the Carrier of Last Resort (COLR). The COLR is required to maintain "Plain Old Telephone Service" for households and businesses across the state by request. AT&T argues that the cost of maintaining these legacy systems keeps them from investing in the modernization of their network. It is currently estimated that without a COLR, more than 580,000 households across the state would be without a telecom option. The North State Planning and Development Collective estimates that if AT&T is successful, the population impact in Butte County alone is over 37,000.[43] Homes with only landlines are often rural. Cell coverage is strongest in urban areas where the infrastructure is built to meet market demands. Here again, we have a disproportionate rural impact.

In the foothills of Butte County where the Camp Fire destroyed the Miocene Canal, a water conveyance system owned by PG&E, households dependent upon that steady stream of water for over a century found themselves dry.[44] PG&E, owner of the conveyance system connected to an inactive hydropower facility, initially publicly committed to the County of Butte an investment to restore the system, but later backed away citing higher than expected permitting and construction costs. Potable water deliveries are the current work-around until a permanent solution with ongoing funding can be found.

Without water, without insurance, without a way to call in or out, rural areas in California are becoming increasingly disconnected from basic services. As we watch these communities burn to the point of total loss, and we see how slowly recovery occurs if at all, with the bulk of subsidized housing going into urban areas, is rural life slowly disappearing and, in the case of fire, vanishing overnight for good?

---

[42] Reported by United Policyholders during a fellowship interview in October of 2025, from data and statistics provided by the California Fair Plan Property Insurance web site, https://www.cfpnet. com/key-statistics-data/.

[43] Data provided to Butte County by the North State Planning and Development Collective with Chico State Enterprises, 2024.

[44] Redirection of PG&E Water Delivery from Miocene Canal to the West Branch of the Feather River Project, CEQA Notice of Exemption, filed with the State of California on April 25, 2019.

4.12 Intact Community

At the 2023 Rural Voices Coalition of California annual conference in Tahoe, I listened to practitioners solving problems in intact communities using intact community infrastructure and amenities. They endeavored to increase middle market housing supply without starting at a 14% housing stock loss. They spoke of prescribed fire as a miracle cure without considering the impact of visible smoke plumes on traumatized children. They talked of building recreation systems through forested areas without contingency plans for wildfire. I envied their unknowing.

At the same time, I heard incredibly beautiful stories about land restoration and economic stewardship that broadened my perspective.[45] I related to the desire to invest resources back into rural communities rather than export them as the only means of economic viability. I appreciated the definition of working lands as economic drivers, of prescribed fire practitioners putting their heads together to overcome insurance obstacles, and of the energy around different approaches to forest management in other states. Thank goodness for these impassioned professionals looking out for our intact communities and land.

But I was failing to see the applicability of their ideas in destroyed communities and felt like an outsider. An inclusive approach would be for groups to acknowledge that their work is wholly contained within intact communities and landscapes, and that destroyed communities and landscapes might need a different approach.

Privilege is a wall disaster recovery professionals hit in local conversations, too. Recently I was invited to a meeting on increasing market rate housing to meet workforce recruitment and retention demands in Chico. Having worked at the Chico Chamber, I've heard local employers struggle with recruiting workers to fill jobs for many years. I know when market rate housing is in short supply and trailing spouse jobs are not available, recruitment falters. These challenges are valid economic development concerns in intact communities, but not immediate disaster recovery concerns in destroyed communities. There are times when economic development goals far surpass what's possible during disaster recovery.

Economic development involves using a toolbox of market data, financial incentives, permit streamlining, government tax credits, political capital, relationships, development assistance, business assistance, and edging out competitors to build economic vitality. In my experience as an economic developer, I can be assisting with a power purchase agreement on behalf of a business one minute, and advising a developer on how to relocate an eagle nesting in a newly purchased warehouse the next. Economic development requires an on-call, all-hands approach in rural California (Fig. 4.2).

Placing sample eligibility criteria side by side makes it clear why destroyed communities cannot compete with intact communities for traditional funding, and why traditional economic development tools may not work after disaster. Relieving wildfire impacted communities from intact eligibility criteria if they experience two

---

[45] Speakers at the conference were procured by Laurel Harkness, Rural Voices of California Coalition, 2023.

Sample Eligibility Criteria

Intact Community	Destroyed Community
Existing Water Source & Supply	No/Damaged Water Source & Supply
Water/Wastewater Connections	No/>50% Water/Wastewater Connections
Roadway System	>50% Damaged/Destroyed Roadway System
Incremental Population Changes	>50% Population Changes
Housing Stock	>50% Housing Stock Loss
Existing Utilities	>50% Utility Damage
Business/Economic Activity	>50% Loss of Business/Economic Activity
Parks/Community Centers	>50% Community Space Loss

Fig. 4.2 Chart showing criteria for defining intact communities versus destroyed communities

or more of the indicators above might open up new funding sources that could increase community viability post-disaster. Federal funding exists for repair and reconstruction, but investing in resilience during recovery will take all available funding sources, particularly those meant to enhance and expand community amenities, infrastructure, economic activity, and housing. I recommend all federal and State funding agencies consider adapting their eligibility criteria to suit both intact and destroyed communities to broaden the bench for recovery, and to increase equity across populations living in both. Otherwise, the benefits of most funding are felt exclusively in communities where disaster is not.

References

Alesch DJ, Siembieda W (2012) Int J Mass Emerg Disasters 30(2)
Johnson, Hayashi (2012) Int J Mass Emerg Disasters 30(2):121–238

Chapter 5
Trauma

When I first joined the County, I dove headlong into an economic developer certification process offered by the California Association for Local Economic Developers (CALED). To fund the first course, I volunteered as an Ambassador for CALED in exchange for waived registration. I devoured the trainings even though my disaster lens got the better of me and I couldn't help but filter every scenario through my recovery perspective.

After two week-long certification courses conducted a year apart, I took the exam. Per the prompt, I prepared an economic development paper and slide deck demonstrating my knowledge and skills. Given the option to prepare a disaster scenario, I selected the real-life example of the Paradise regional sewer connection project I'd worked on in the Town that would significantly increase housing density and economic capacity after the Camp Fire.

In public documents, the sewer connection is described as the single most impactful economic development project in the Town's recovery. The Paradise Sewer Project Final Program Environmental Impact Report (PEIR) Executive Summary prepared for the Town by HDR, Inc., explains that historically, "reliance on septic systems has resulted in two areas of concern: environmental impacts and economic impediment."[1] The study goes on to explain, "the lack of a sewer system has suppressed the development of a sustainable business community by limiting the size and types of businesses that can affordably operate in the community."[2]

The PEIR references a feasibility study conducted prior to the Camp Fire by Bennett Engineering for the Town that showed projected benefits of a sewer project would include an additional 161 jobs, an additional $12.8M in sales, and a 5–13% increase in property values for parcels within the sewer district. When the Town re-evaluated the project in late 2019, just prior to my arrival on staff, they found as

---

[1] Paradise Sewer Project Final Program Environmental Impact Report (PEIR) Executive Summary prepared for the Town by HDR, Inc., page xix, https://paradisesewer.com/application/files/9717/2046/7145/EXEC-Summary_FINAL.pdf.

[2] PEIR, page xix.

K. Simmons, *Three Fire Mountains*, https://doi.org/10.1007/978-3-032-17343-0_5

the PEIR explains, "businesses have been severely constrained due to their septic system discharge exceeding the available capacity of the land, while new businesses are often forced to open elsewhere due to the limitations placed on them to operate with an on-site septic system."[3] Therefore, the second goal of the project just behind accommodating regrowth in the Town post-disaster is to generate economic recovery.[4]

This one wastewater infrastructure project proposing to connect 1,100 parcels currently reliant upon septic to a Sewer Service Area is recognized as the primary means for restoring and increasing economic vitality in the Town's downtown core and in primary commercial and mixed-use districts. Shifting parcels from septic to sewer allows for increased water usage onsite which reduces lot sizes for housing, creating greater density, and allows for higher water usage businesses like restaurants to rebuild.

In my exam conducted virtually by CALED, I explained how the project is improving economic prospects for an entire community by measurably increasing housing, business capacity, population, and tax revenue. Additionally, establishing higher density housing in the center of town situates more of the population on the primary evacuation routes. I explained how I as an economic developer would play a pivotal role in teasing out the economic benefits of this project and communicating them to the many local and State agencies involved in permitting and funding the project. I had to "sell" the project on its economic value to the decision-makers.

During the exam, I felt confident describing the economic impact of this example because I'd once been responsible for setting up the joint Town-City meetings to develop the memorandum of understanding governing the project. The MOU outlined how increased capacity in the Chico Wastewater Pollution Control Plant due to the regional pipe from Paradise would open up housing opportunities in Chico where many fire survivors had relocated, meeting rising demand. Win–win–win for Paradise, Chico, and the region in great need of housing as the primary driver for economic recovery and regrowth following disaster.

As soon as the Q&A portion of the exam started, however, I could hear the resistance—the examiners' inability to leap from an intact community mindset to a destroyed one. I went from confident to defensive to defeated in a matter of seconds as their doubts overtook my certainty. I could tell they could neither hear nor see the enormity of the project's benefit on the decimated economy in Paradise. My heart broke a little bit more for the town.

A few weeks later, the results arrived and I knew CALED's decision before I opened the letter. They failed my exam citing that while they understood we'd had a "large fire," I had failed to demonstrate my knowledge of economic development. My example didn't meet any of the standard techniques taught in the training sessions; it was simply a public works project.

In community recovery planning and in federal funding priorities, infrastructure supporting housing is often recognized as the single greatest necessity for recovery,

---

[3] PEIR, page xxiii.

[4] PEIR, page xxiii.

which is why economic developers in recovery roles put their focus here. In the Town of Paradise pre-fire, 95% of the jobs were local-serving which means without population, few jobs could be restored.[5] Without housing there could be no population, and without infrastructure there could be no housing.

Hence, economic recovery—which is economic development following disaster—starts with infrastructure. Well, let's be honest, economic recovery starts with response, then fire suppression repair, then debris and tree removal, then planning, then infrastructure. Economic developers are involved in all of these steps, analyzing the economic gains of each, aiding in prioritization to the extent possible, and securing funding. Take debris removal. If debris is not removed, reconstruction cannot occur. I like to say disaster recovery is the only field that requires professionals to answer which comes first: the chicken or the egg.

I'm not sure what upset me more about the failed exam, the examiners' dismissive tone toward my skills and experience, or their reference to the state's deadliest, most destructive fire to date as a "large fire." Both cut me to the core. What I read between the lines and later rebutted in a letter back to the examiners was what I perceived as privilege. Sure, this is not what an economic developer would do in an intact community with all the tools, eligibility, and amenities at hand. With a thriving business district and neighborhoods, an economic developer would not be developing an MOU between agencies for sewer flow exchanges. No, that economic developer would be pulling the financing tool out of their toolbox and using it on a single under-developed lot to entice a single project, creating a handful of jobs and a portion of a percentage point in sales tax revenue. My colleague in the North State used this example on his exam and passed.

In their reply back to my rebuttal they doubled-down, offering me mentoring so I could learn the real ropes of economic development and try again on the exam. They offered the support of an economic developer working in a middle-income, mid-sized suburb of the Bay Area who, to his credit, guided me to stick with what economic development looks like during recovery with the addition of a few key words.

I feel haunted to this day by the exam experience, and rage over the irony of being shut out of my industry association during the hardest chapter of my career, when that association itself cannot guide us after the worst has happened. Watching colleagues in other disciplines such as social services and community planning, I see their associations honoring the depth they can now bring to their conferences having worked through the Camp Fire. In the case of CALED, the examiners were as unyielding with their comments the second time, but certified me by a tiny margin in my "pass" letter. A few months later, they announced the certification was only good for three years, something not communicated to us during the process, and I had a hard time not taking this personally.

I dropped my membership to CALED and unsubscribed after a few more email exchanges with their executive left me even colder. I kept the certification initials,

---

[5] Camp Fire Regional Economic Impact Analysis, Final Report, January 2021, Economic & Planning Systems, Inc., and Industrial Economics Incorporated, p. 54.

"ACE," on the signature line of my email for a couple of months then deleted them as well. There's no pride in receiving the certification when the process was so painful. I thought playing by the rules and trying again might be a healing process for me, but it undercut the tenuous faith and trust I'd built in my newly required skill set. CALED didn't engage me in a relationship at that time, nor did they reach out to me personally to listen and understand despite my efforts to communicate. It's possible the emotional build-up on my side wasn't visible enough for them to respond, or it was too visible. Trauma can be difficult to navigate in a professional email. They published my name in their newsletter when I passed as if the celebration was mutual. I saw it and thought, "how dare they," then tossed it in the office recycling bin.

What I took away from this experience is that the traditional definition of economic development assumes the presence of an intact community, period. It doesn't even factor in the risk of disaster when evaluating tools for the job. Similarly, the agencies that fund economic development assume the underlying assets of an intact community are within arm's reach. A brownfield lot can be transformed into a complimentary use with site remediation, proper zoning, financial incentives, the right developer, and political will—which is a bit like disaster recovery within an intact community. What about when an entire community is a brownfield, filled with hazards and waste above and below ground following wildfire? Do we not consider reconstruction economic development? Economic development according to CALED's review of my exam is a blue-sky endeavor.

In traditional economic development, there is no consideration of destroyed infrastructure at scale, the presence of fire debris and hazardous trees, a pending settlement determination holding up escrow and locking destroyed lots from redevelopment, the un-insurability of lots due to high fire severity risk modeling, and community-wide trauma and perception about risk. In early recovery, we have to address these conditions before jumping into traditional economic development.

In a post-fire situation, I consider disaster recovery the precursor to economic development because it lays the foundation for economic recovery. And why not consider the two disciplines part of one long continuum? You have to start somewhere after fire and if you pull out what you learned in your economic development training as the smoke clears, you're in for a shock.

In the Town my title was Disaster Recovery Director. In the County my title is Deputy Chief Administrative Officer. The intent behind both is recovery. The realities of both require economic development. It pains me to think this way, but when more communities in California are destroyed, the traditional definition of economic development will expand to include the tools of disaster recovery, all the way down to evaluating the economic benefits of the sanitation pipes that make reconstruction possible. The Town may be the first community taking economic development subterranean following catastrophic wildfire, but they're certainly not the last.

Professionals who have spent their entire careers in economic development may find themselves in disaster recovery roles overnight, not by choice but by circumstance. If they do, I hope they reach out to other communities grappling with the same challenges to learn from their experiences—because that's the only way to alleviate the loneliness of losing a job to fire, and finding the one that replaces it is nothing like it used to be. Professionals working in destroyed communities find kindship after fire which is why it's never a burden to hear from LA and Maui recovery staff.

I spent two decades working in local, state, and international membership associations like CALED prior to entering the field of disaster recovery. The goal of professional associations is to find and share common ground for networking, education, and opportunity. I regret to think now that we acted exclusively, but maybe that's the nature of associations by design. Any organization using words like "joining" and "member" is assuming an in-group and an out-group. I spent my whole career in an in-group until I transitioned to the disaster recovery field where there is no group at all.

When I prepared my first exam materials, I relived the fire and my state of mind during those early project days. I went back and researched the details so I could approximate the measurable economic value. When I presented my materials to the examiners, I was shaky and emotional like I am every single time I talk about the Camp Fire with a new group. I allowed the exam to take an emotional toll on me because of how significant I thought the project was to share. The failure felt like a rejection of my hard-earned skillset and the personal price I pay for doing work of this kind.

My friend Caryn Albrecht turned it around for me. She said consider the failure proof that you're doing something specialized, not mainstream. Be a unicorn.

5.1 Trauma-Informed Approach

Through this writing I've started to understand what my mother needed from her insurance agent when she wanted him to care about her losses. She needed a trauma-informed approach. This practice is used in many fields and I've come to appreciate its applicability to working with fire survivors and with myself.

In my definition, taking a trauma-informed approach with fire survivors is meant to limit the amount of re-traumatization caused by reliving the events of the fire in pursuit of recovery. It assumes they've lived through dangerous and/or heartbreaking circumstances and have experienced trauma as a result.

We made many calls to enroll households in the hazard tree removal program after the Camp Fire. Sometimes we'd call several numbers before reaching a person because of how many times survivors had moved since the fire. When we finally had someone on the line, we learned we couldn't jump right into business without first hearing their fire story. If we had two minutes of business to discuss it was preceded by what happened to them on November 8th and during the fire's aftermath.

Sometimes the stories were frame-by-frame from the moment they learned something wasn't right, and sometimes that part of the story was brief followed by their husband's long decline. Or, they talked about the health issues they've had since the fire, or how making decisions about the property drove a wedge between them and their adult children. The stories would last a minimum of an hour and well before

I knew about secondary trauma, or what a trauma-informed approach was, I knew the only way to get to business was through the trauma. Tree removal itself was re-traumatizing because it wouldn't exist without the fire.

In disaster recovery, every phone call you make is about the disaster. After a while I realized I couldn't absorb more than four stories a day before my own well-being faltered. Once I heard the fourth story I'd have to switch to another project. The Town Manager graciously allowed me to hire what I called "Tree Advocates" to help with the phone calls because by the time I was hired we had thousands of households still unenrolled. The Tree Advocates were amazing ladies who seemed to have endless capacity for devastating stories. I'm not sure now if I found the work challenging because it was so re-traumatizing for me, or if I lack whatever special quality they had. The Tree Advocates were miracle workers.

I've observed a few other miracle workers during my time in recovery outside of the first responders. Support dogs who visit incident command centers comfort firefighters and personnel during times of stress. I had the opportunity to meet a few of these at the Dixie Fire Incident Command Center, and I've seen them at our County offices as well. They're not quite the miracle workers on the fire line, of course, but they are cute and certainly lowered my blood pressure.

I believe the hazard tree removal program opened my eyes to the complications of both experiencing and recovering from a fire—healing inevitably stalled by re-traumatization. For survivors, there was nothing in their lives that did not relate back to the Camp Fire in those early days, months, years. The disaster interrupted everything—jobs, homes, grocery stores, parks, friend's homes, children's activities, relationships. Sometimes it wasn't the fire that broke them but its aftermath. This is how I came to know that fire is only one layer of destruction and hardship, the consequences such as health complications, legal troubles, family strain, hospitalizations and death, and that ongoing bewilderment and feeling of isolation from a world that is moving on is sometimes much, much harder on a person than escaping from the flames.[6]

I've needed a trauma-informed approach myself over the years. I certainly needed it from CALED, and I need it from my consultants who puzzle when I ask for the time and date stamp of their data. "Why does it matter when this household data was collected?" they ask as they fly me over Google Earth looking at pinpoints on the map, and I will have to explain that the entire area they're studying burned in 2018 or 2020 or 2024 and the data they're using may not reflect the current population depending on when it was collected.[7] There is a difference between the number of households in Butte County prior to and after our recent fires. If the data is from March of 2018, I'm not going to want to use it to understand Camp Fire realities. If the data is from January of 2020 and includes communities impacted by the North

---

[6] Numerous observations and anecdotes from family, friends, and working in disaster recovery following the Camp Fire.

[7] Observations and meetings with consultants during planning processes following the Camp Fire and North Complex Fire.

Complex Fire later that summer, it's not valid. Collection dates matter after disaster if you're using data for recovery.

If there's still some confusion, because that's the hard shell of intact community blindness, I will pull up fire footprint maps and show consultants what I'm talking about. If we're still not getting anywhere, I will share imagery of the post-fire apocalypse to demonstrate a level of destruction otherwise hard to explain. To bring consultants to the current moment, I will share photos of the recovery and the rebuild, and read off the number of building permits and certificates of occupancy issued in the burn scar to date.

Sometimes the light will dawn and there will be a connection between what they've seen on the news and the reality of our community a few years into recovery. I will explain my personal theory that an intact community iterates 5% every 30 years, and a destroyed community iterates 50% every 6 months. I try to explain the rate of change when a community is whole, when it is destroyed, and while it begins to rebuild. All are very different.

If the project requires us to go deeper, like if we're preparing for community outreach in impacted areas, I talk about what happens to schools, even those still standing, when houses are destroyed and families move away. I share the basics, the obvious, the nearly impossible to comprehend. Sometimes we get somewhere and I will see the lightbulb click on. The consultant or team will pause and sit back. On one call an apology came quickly—which is never what I'm after but usually follows my 'Destroyed Community 101' lecture. Another called me in tears a few days later. Some consultants leave our calls and watch news pieces or documentaries on the Camp Fire by their choice, never by my recommendation, and be changed. Destroyed community reality sets in.

For example, as we were finalizing our Butte County Broadband Planning and Feasibility Study with Tilson Technologies in early 2024, I had a feeling the unserved/underserved areas of our community are also the hardest hit by wildfires. I shared this hunch with our consultants and asked them to layer the Camp Fire, North Complex Fire, and Dixie Fire maps over the broadband service maps. To accomplish this, I provided publicly available shapefiles of our burn scars, and the consultants obtained the service maps from the State. Sure enough, the layered maps told a troubling story: the areas least served by broadband are also among the areas most recently hit by fire.[8] For emergency communications planning, it's essential to look at levels of service in high-risk areas, and consider traditional and alternative means for increasing uptake and redundancy to avoid system failures (Fig. 5.1).

I admit that on some calls with consultants I am fatigued by the effort it takes to explain our fires, and I ask them to do their homework and brainstorm tools and solutions before calling me back. There is plenty of publicly available information on our fires that can orient consultants in ways that are not taxing to local staff. Some consultants do this on their own. One consultant we worked with in early recovery pitched a new business model to his boss once he understood destroyed communities need an entirely different approach.

---

[8] See Fig. 5.1.

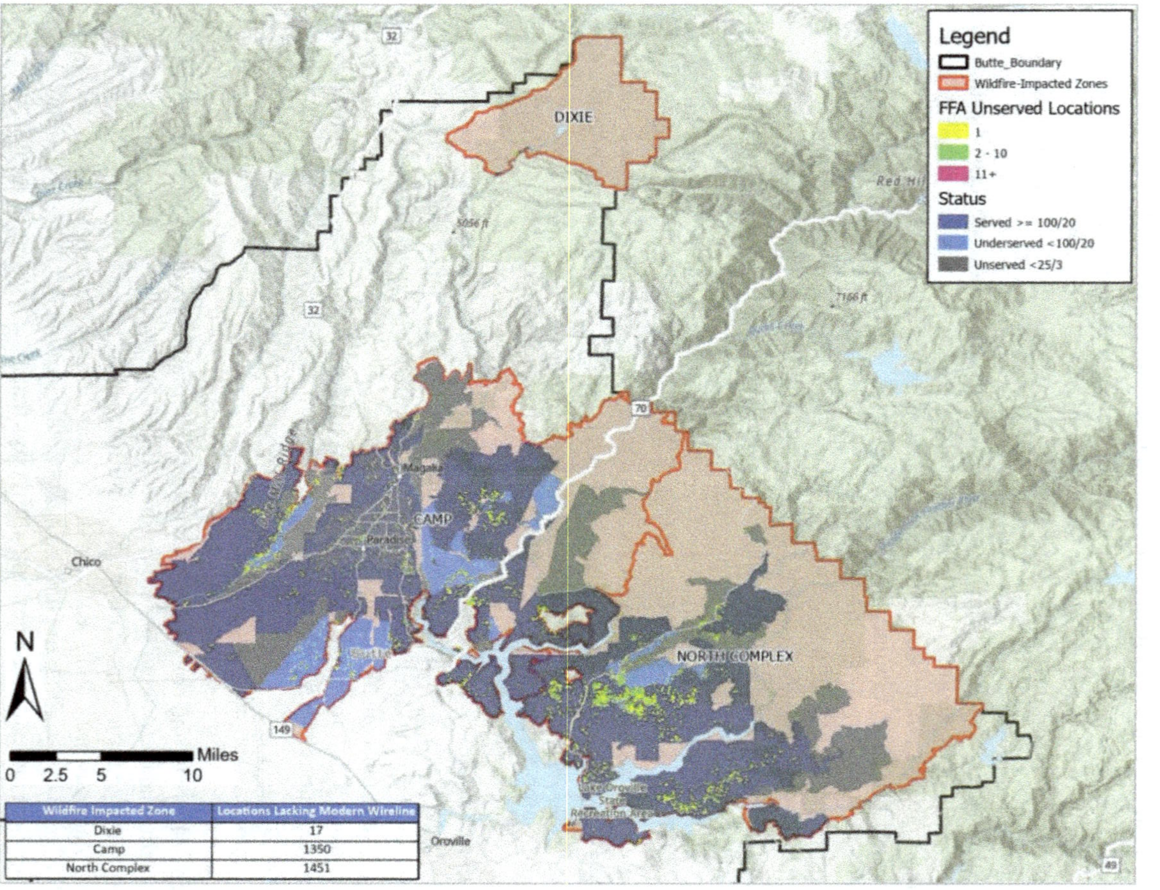

Fig. 5.1 Specialized map on page 12 of the Butte County Broadband Planning and Feasibility Study showing broadband access in the Camp, Dixie, and North Complex burn scars within Butte County

A few years after the Camp Fire, a local non-profit invited national consultants to Butte County for a strategic planning session. To relate to the audience, I suppose, the consultants prepared a segment on disaster recovery and we sat through dozens of photos of community destruction. They played a 10-minute video on a community's recovery from a tornado which took us from harrowing imagery to quotes like, "this was the best thing that could have happened to us." In this case, the tornado hadn't caused any fatalities, and I marveled at how when that is the case, communities can turn disaster into a gift. When there are fatalities, no community will ever be grateful for a disaster.

To this day, I remember visibly shaking during the presentation and ultimately having to excuse myself. I thought about raising my hand or privately speaking with the consultants to explain the impact of disaster imagery on traumatized audiences. I wanted to point out the differences between the best and worst thing that's ever happened to a community…but I didn't…because I thought my opinion wasn't important. I kept my thoughts and feelings to myself then, but in the years since, I've learned to let out little bits and pieces of those lessons so I don't have regrets, and so consultants can consider the ways in which traumatized people might be harmed by what they're sharing about disasters.

I will never know where consultants take information from disaster recovery staff—they may forget immediately. My deep, perhaps idealistic hope is that they will handle destroyed communities with greater sensitivity in the future. I am starting to say the words, "trauma-informed approach," with greater frequency to cut to the chase. I told one consultant her virtual meeting background depicting a raging wildfire was perhaps not the most trauma-informed approach for engaging with disaster recovery professionals on wildfire protection planning. At minimum, I hope consultants understand there's a better way to jump into a working relationship with a community in recovery.

My suggestions for working with disaster-impacted communities with a goal of reducing re-traumatization:

1. Determine if a major disaster has occurred in the jurisdiction within the last 10 years.
2. If yes, assume the presence of trauma and consider the following:

   (a) Research all publicly available information on the disaster—media coverage, After Action and Corrective Action Reports, data reports (DINS, etc.), online images, other credible sources. This research is meant to create consultant awareness so that the client does not have to re-experience the disaster by describing it.
   (b) Research common signs of trauma for situational awareness; do not discuss trauma with the client unless invited. It's possible the client did not experience the disaster, but the community-at-large did.
   (c) Recognize every plan developed within 10 years of a major disaster will likely contain elements of recovery, regardless if it is called a recovery plan or not.

(d) Date-stamp and time-stamp all data to give the client the opportunity to use it in a pre-disaster and/or post-disaster context. Understand the timeframe of data the client is asking for, and anticipate that major swings in data may be present at the time of data collection.

(e) If the project requires mapping, request shape files of the fire footprint to show the client if the work performed or proposed occurs within the burn scar. Know what a "burn scar" is if working in wildfire impacted areas.

(f) Review the destroyed community criteria presented here. If the community meets the criteria, do not assume the presence/availability of intact community resources and amenities, nor grant eligibility for traditional funding sources.

(g) Avoid disaster comparison in conversation with the client. It may be appropriate to relate to the client, but not engage in comparison. No two disasters are the same in terms of losses of people and property, and proximity.

(h) Acknowledge the consultant's "outsider" perspective, and ask if there are additional sensitivities to offer the community during outreach.

(i) If the client and/or community offers their disaster story, take in the additional context and ask if/how the information should be included in the project. Recognize if the plan or project is for recovery, and the focus is on the disaster, many survivors begin by sharing their disaster story. In community outreach, make plenty of time for disaster stories, and prepare staff for any secondary trauma effects.

(j) Consultants should take care of their own wellness and find resources to support working in, on, and around disaster.

My advice if you know a community has experienced disaster of any size is: do your homework to understand the magnitude and impacts, assume the presence of trauma particularly if you are working with survivors, and be aware of your actions to minimize re-traumatization.

5.2 Legal Action

I have learned lawsuits follow fire. PG&E pled guilty to 84 felony counts of involuntary manslaughter after the Camp Fire.[9] I happened to pass by the courthouse in Chico on the morning of the verdict where media vans were broadcasting by sunrise. To say a pall hung over the building is putting it lightly. The community was in mourning.

The Fire Victim Trust is the third-party administrator entrusted to make determinations and settle claims with fire survivors using bankruptcy proceeds from PG&E. Claims can be filed for economic and non-economic damages and can account for

---

[9] As PG&E Pleads Guilty to 84 Deaths in Camp Fire, Report Says It Put Profits Over Safety, Michael Liedtke, Janie Har, The Associated Press, KQED, June 16, 2020, https://www.kqed.org/news/118 24596/pge-pleads-guilty-to-84-deaths-in-wildfire-that-destroyed-paradise.

lost wages, business losses, personal injury, emotional distress, and death.[10] A local business owner called me recently saying she is trying to reopen her business five years after the fire and is still negotiating her PG&E settlement determination from the Trust, which is insufficient to rebuild.

Settlements from the Fire Victim Trust for the Camp Fire trickle out in percentage payments years after disaster, with sizable chunks removed for attorney fees.[11] What's left is taxable unless legislators exempt disaster settlement payments from State and federal tax requirements. It seems lump sums would accelerate recovery, but that's not how these payments work.

I interviewed two Town staff members for the fellowship, both of whom I worked with at the Town. Colette Curtis still works in recovery and the other, since retired, worked in housing at the time of the interview. I have deep admiration for their fortitude. I asked what data they use to determine if their housing urgency ordinance should be extended. The County was grappling with this question when I submitted my fellowship application to Stanford thinking research around data and evidence could help. In the case of the Town's urgency ordinance, as it turns out, local data may not matter, which goes to show that policies related to the same fire in two different jurisdictions may need different approaches. Urgency ordinances are not one-size-fits-all when circumstances vary across the fire footprint.

After the Camp Fire, the Town and the County approved a series of urgency ordinances within their respective municipal codes to allow for temporary uses of land to address urgent needs, reduce health and safety risks, and permit uncommon circumstances following disaster. The County consolidated its urgency ordinances for the unincorporated area into Chapter 53 for the Camp Fire, Chapter 54 for the North Complex Fire, and Chapter 56 for the Park Fire in 2024.[12] County staff is now looking at codifying these ordinances so there's a template for the next fire.

Chapter 53 was established in late 2018 and expired on December 31, 2023. Chapter 54 was established in 2020 and extended until mid-2024. Chapter 56 was established in mid-2024 and is ongoing as of this writing. These chapters allow, among other things, for property owners to live on their lots in temporary housing without active building permits. The idea behind the urgency ordinance is to conditionally permit RVs as temporary housing on burned lots during the time it takes property owners to plan, finance, permit, and execute their rebuilds. In the County, the urgency ordinance following the Camp Fire was extended a few times for a total of just over five years.

The Town Council, like the County Board of Supervisors, reviews and discusses the need for each urgency ordinance as it approaches expiration, as they have decision-making authority over ordinances. The Town Council extended their

---

[10] Fire Victim Trust established in 2020, https://www.firevictimtrust.com/.

[11] Self-reported by fire survivors in numerous anecdotes and reported through a public survey conducted by Butte County to determine the rate of pay-outs and responsiveness of the Fire Victim Trust, 2023–2024.

[12] Butte County Municipal Code is available at https://library.municode.com/ca/butte_county/codes/code_of_ordinances.

urgency ordinance several times in 6-month to 1-year increments, as has the County Board for each of its recovery chapters.

Interestingly, within a few months of the Camp Fire, the City of Chico set an expiration date for their RV allowance well after the 5-year mark and did not review it again. I suspect the fears of inhibiting or stalling recovery in intact communities are different than they are in destroyed communities. Whereas a destroyed community may worry that extending temporary housing may stall its return to intactness, the City of Chico did not appear to fear the loss of its intact status by allowing RVs to park in front yards and on construction sites with the right permits. The City of Chico addressed the housing crisis they did not cause, and met needs they otherwise couldn't meet with existing housing stock, by allowing temporary units to fit within built-out neighborhoods for an extended period of time.

Town and County concerns resulting from direct fire impacts are different than the City of Chico's concerns resulting from indirect impacts. From my observation, the balancing act comes from permitting temporary uses that allow people to transition to recovery while not prolonging destroyed-community living conditions. In some cases, RVs are in compliance with the regulations and are properly permitted with conditional use permits that certify the presence of hook-ups to electrical, water, and wastewater systems. In other cases, RVs are not permitted and may be leaking raw sewage onto the ground and into neighboring lots, or may be powered by unpermitted electrical cabling or noisy generators. In one destroyed neighborhood I saw a hose running from a weedy ditch into the window of an RV.

RVs on private land are occupied by property owners, renters, and people without explicit permission to be there. Some renters bought vacant properties after the fire hoping to build, but found the process far too expensive and remain in RVs. Today, it's not uncommon for large lots to have more than one permitted or unpermitted RV parked onsite. I'll never forget one particular day working in the Town when a grass fire was ignited by small equipment on a weed-filled lot, and minutes later a report came in that an RV was in flames. During outreach on one lot with an occupied RV, we came across a plywood outhouse with a bucket inside.

Preparing for each round of urgency ordinance decisions, Town staff prepare an analysis of temporary housing by quantifying the number of conditional use permits issued for RVs, building permits applied for and issued for single-family and multi-family housing, and relevant code enforcement cases related to unpermitted RVs and other health and safety violations. Town staff monitor living conditions closely, making regular contact with unpermitted RV dwellers to understand each and every barrier to reconstruction and/or to encourage permanent housing programs and options. Even with similar outreach in the County, a handful of residents recently said at a Board meeting they will never vacate their RVs because rebuilding is prohibitively expensive.

As Town staff shared with me, prior to presenting this data to their Council for the latest round of extension discussions, HCD said on a routine program call that if the Town allowed their urgency ordinance to expire, the State might take legal action to recapture their disaster recovery funding (CDBG-DR). The warning didn't just include funding for housing but for infrastructure, facilities, planning, public services, and mitigation, for a total of over \$200M.[13]

As precedent, Town staff explained, HCD described lawsuits against jurisdictions across the country citing violations of the federal Fair Housing Act. The Town Attorney found the warning credible and advised the Town Council to extend their urgency ordinance which they did.

Jurisdictions must comply with federal regulations per their funding agreements. It is an obligation we take very seriously. However, I was surprised to learn this was how HCD chose to convey their compliance concerns to the Town. I had a general sense of the data the Town used for these discussions which I'd been involved in a few years prior, and knew the work several departments put into gathering and analyzing that data for presentation to the Council. The Town does not make these—nor any—recovery decisions lightly.

Delivery aside, the timing of the warning appears to have coincided with HCD winding down the owner-occupied reconstruction program in the Camp Fire burn scar which offered forgivable loans to qualifying households for reconstruction in the Town and unincorporated areas.

Expiration of the Town's urgency ordinance would likely affect some of the property owners HCD's program-of-last-resort had failed to assist for any number of reasons: eligibility, trust, delays. It appears HCD hinted that the Town should bear responsibility for this lack of programmatic success, risking funding that was supporting the entire community's recovery. Town staff told me they'd planned to recommend renewing the urgency ordinance anyway, because local data indicated it was necessary.

This breach of trust during an already challenging recovery process stems from a lack of coordination and communication between federal and State agencies assisting local governments with recovery. Urgency ordinance deadlines are public information, and can be tracked by State and local governments while grants are underway. Any concern about expiration should be discussed well in advance, and handled with enough time for the jurisdiction and State to determine any possible violations, so the right information can be provided at the time an extension is considered by elected officials. This coordination goes well beyond collecting data to understand needs and impacts before informing local policy decisions.

In the Town's case, whether HCD's comment about recapturing funding was made casually or with real intent, it lacks a spirit of partnership between agencies meeting at least monthly on dozens of recovery programs. There's no reason information pertaining to federal compliance should be withheld or delivered without context or appropriate time to react. Certainly, if a violation looks imminent, the local jurisdiction should be notified by any means possible. By contrast, a coordinated system for recovery—versus regular check-ins on various programs—would catch such pitfalls without imposing the terror of losing critical recovery funding.

To HCD's credit, several programs operate with a growth mindset, and representatives listen to Town and County staff with a sincere desire to learn about rural

---

[13] Fellowship interview, Town of Paradise staff, 2024.

roadblocks during recovery. Maziar Movassaghi, the very first person who appeared on screen from HCD during our COVID-era disaster recovery calls, still encourages me to give him "magic wand" solutions to the challenges our agencies face in recovery, so they may anticipate and adapt their programs to local needs going forward. I regularly take him up on this offer and partnered with HCD on a Park Fire housing recovery survey when it was clear a federal declaration was unlikely.[14]

5.3 Prism Effect

The Chief Administrative Officer for Butte County, Andy Pickett, and I exchanged words describing the ways in which disaster recovery has opened our eyes. I used the word 'disillusionment' to describe my experience. I believe what has broken my spirit more than witnessing catastrophe is realizing "the system" where adults in agencies "work together" to ensure "the right thing" happens is not a system at all. There is no invisible construct in disaster recovery ensuring cooperation and trust. There are incredibly wonderful people doing their best in all levels of government, but the sum total of their potential as employees far outweighs the capabilities of those agencies.

My growing disillusionment says more about me than the world around me, though. My lifelong inherent trust likely comes from the fact that I was not overtly harmed by agencies when I was a child. My privileged perspective has collided with real world realities in disaster recovery, and I would describe the reckoning as a literal wound on my soul. Expecting an agency made up of thousands of individuals to be perfect is ludicrous when expecting even one human to be perfect is unreasonable. I don't expect perfection, but I do expect something closer to the right thing.

Working in local government, I have come to see public perception as a composite of millions of tiny details that can't be known or changed by one person, one agency, or combination thereof. This is unsettling for someone who loves a bit of control and predictability. County government continues to teach me not to come to conclusions too quickly or at all. I hear a brief report-out in a meeting and think I know the angle and end-result. Turns out, as soon as our Chief Administrative Officer describes the situation in more detail—which I call pulling back the curtain—eighteen other dimensions appear that I hadn't considered or foreseen.

I call this the prism effect of government. Light doesn't just pass through glass to illuminate the other side, it hits a prism and fractures into a billion colorful rays bouncing off of every 3-D surface in every imaginable direction. I try to remind myself of the prism effect when I get too attached to the "truths" my disillusionment speaks.

As incensed as I can get with what I observe and learn throughout disaster recovery, I remember the prism effect. As much as I want to understand every angle of every

---

[14] Park and Borel Fire Survey conducted by the CA Department of Housing and Community Development, 2024–2025.

prism, I know the pursuit of that will leave me with no time for anything else. And even if I had the time, the whole truth is unknowable because light travels faster than the eye can move and certainly reaches far beyond what the eye can see.

5.4 Government Continuity

A few months ago in a staff meeting, I was inspired to draw a picture on the white board in our conference room to welcome a new member to the team.

I drew a big box at the top and labeled it universe. Below it a smaller box labeled galaxy. Below that a row of small boxes labeled solar systems. Then, directly below those solar systems, planets. I explained that the Board of Supervisors sits in the universe privy to all there is. Our Chief Administrative Officer sits in a galaxy just below the Board with a view of the universe and the solar systems below.

Within the solar systems are deputies and liaisons who connect the Administration Department to the many other departments (planets) within the County: Auditor-Controller, Clerk Recorder, Treasurer-Tax Collector, Sheriff's Office, Butte Fire Department, District Attorney, Assessor, and so on. Those planets are operated by the County's elected and appointed Department Heads and staff.

I explained that a planet represents a whole world, just like the earth—an entire complex ecosystem. It can feel like everything to know, be, and do is right there on that planet. But one only has to look out into the night sky to see the broader solar system, the galaxies beyond, and our great universe, to know each planet is but one tiny object in a sea of infinity.

I believe great leadership comes from knowing and compensating for weaknesses better than mastering strengths. Similarly, I believe success in local government comes not from understanding an individual planet, but the relationship of that planet to all others.

Local government is the space between each planet, the gravitational pull holding them in place: the continuity of operations after disaster, the constant coordination between departments to comply with regulations, the suspense of quick judgement in deference to the prism effect. Local government is a container for the widest variety of skills and purpose, meeting vast community needs.

At the end of the whiteboard meeting, one long-time County employee jumped up and drew a rocket ship heading straight for a planet. "That's Cal OES coming for Emergency Management," she laughed.[15] Even when a rocket ship hits a planet, and staff spends days reeling from the shock, the planet's orbit does not change because it's held in place by everything else. Gravity keeps the planets rotating around each other in neat, orderly lines, the sun illuminates from its corner of the solar system, and the universe carries on.

---

[15] Observed from Butte County Emergency Management staff, 2024.

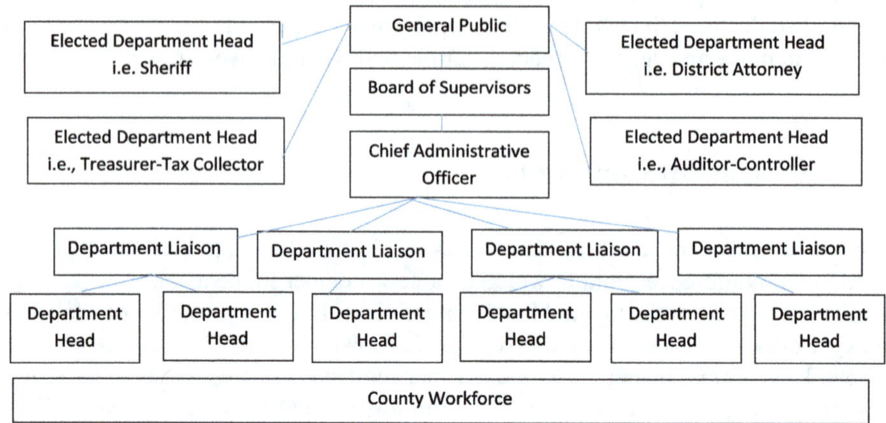

In 2017, the Oroville Dam Spillway Incident hit the County forcing the evacuation of 188,000 people, widely reported as one of the largest evacuations in state history.[16] In 2018, the Camp Fire hit the County. In 2020, the North Complex Fire hit the County, the pandemic in 2020, the Dixie Fire in 2021, the Thompson Fire and the Park Fire in 2024… Who knows what rocket ships are coming next? And still the County functions.

---

[16] 2017 February Oroville Spillway Incident, Cal OES After Action and Corrective Action Report, https://www.caloes.ca.gov/wp-content/uploads/Preparedness/Documents/2017-Oroville-Dam-Incident-AAR-Approved.pdf.

Chapter 6
Industry in Design

Clay Kerchof with HCD explained in our fellowship interview that, historically, 90% of the annual disaster impacts in the nation come from flooding.[1] Given that, it seems reasonable to assume the Federal Emergency Management Agency knows flood better than fire. FEMA pushes timelines suited for flood by rushing wildfire response and recovery.

Recently, Cal OES interviewed our development services staff about the feasibility of ending FEMA funding for additional building and planning staff 180 days after disaster. We explained we are five years out from the Camp Fire and still staffed well beyond pre-fire days. It takes many years for insurance and settlement claims to be determined and paid out before individuals know what resources they have to rebuild. Six months after fire, the burn scar is still ashen and fires smoldering inside of tree trunks have just gone out. My parents' home did not burn but they were unable to move back in until well after the 4-month mark post-Camp Fire due to restricted access to the burn scar and smoke remediation. Five years out and many fallen trees later, they have just begun to rebuild their garage.

I understand from talks with Cal OES early in my time with the Town of Paradise that flood recovery can occur more quickly than fire recovery because flood response is a well-oiled machine and the destruction is different.[2] After flood, it is feasible to imagine temporary housing units moving into the disaster site within a few weeks, and homes under construction a few months after that. To think rebuild permits for more than 13,000 single-family homes in rural, remote disadvantaged communities could be pulled within 6 months of fire is incomprehensible.

FEMA housing trailers for fire survivors did not arrive in our community for more than a year after the Camp Fire, and because they were not compliant with Wildland Urban Interface standards, they had to be placed in valley communities

---

[1] Fellowship interview, Clay Kerchof, CA Department of Housing and Community Development, 2024.

[2] Comments and observations from State staff during Camp Fire Hazardous Tree Removal Operations.

K. Simmons, *Three Fire Mountains*, https://doi.org/10.1007/978-3-032-17343-0_6

rather than on survivors' properties in the foothills.[3] For both the Camp Fire and North Complex Fire, I worked with staff to build cases for time extensions on FEMA Individual Assistance and we were successful. How can a displaced person plan for reconstruction when they are couch surfing, living in an RV, out of state with a relative, or on the streets for more than a year?

During hazardous tree removal from the Camp Fire, the federal government put enormous pressure on the State and local governments to speed things up. When the State finally released the Request for Proposals and selected contractors for the program, there were numerous protests on the award and the solicitation had to be flown again. When the contracts were finally signed, there were immediate change order requests.[4] The first tree was felled around the 2nd anniversary of the Camp Fire in November of 2020. I gathered with hundreds of people to watch the first tree fall, all of us huddled under umbrellas on that rainy autumn day. To expect a community to rebuild within six months after fire when eligible hazardous trees are not felled for another year and a half, finally clearing lots for safe, feasible reconstruction…again, incomprehensible.

Clay Kerchof calls wildfire recovery an "industry in design," and helped me see where the differences between fire recovery and flood recovery stem from.[5] Take the many volunteer groups who arrive in the aftermath of an urban flood to rebuild housing. They are in need of their own housing which is often accommodated in churches and other intact spaces where sewage capacity is not an issue. We tried this in the Town of Paradise in early recovery, but churches still standing did not have the septic capacity for habitation at any scale. In the mountainous wildland urban interface, complying with building code regulations for workforce housing is far more complex than in valleys prone to flooding.

Cal OES staff explained that the increasing frequency and scale of major disasters is unprecedented. They attributed a lot of the downstream complexity in implementing recovery programs to dynamics at the federal level. Specialized knowledge is needed locally to tap into federal funding without creating new risk. Cal OES staff acknowledged that a lot of recovery resources go underutilized for this reason. There are efforts to create navigators to assist, but those navigators must anticipate rural challenges better than local staff to be effective in resolving issues quickly.

The fastest housing reconstructed after disaster is commonly built by adequately insured homeowners. Clay explained adequate insurance is the single most important factor for a timely rebuild.[6] As insurance becomes more expensive and harder to access, the lack becomes a more pressing and complicated challenge for recovery.

---

[3] Learned from Butte County staff who worked on trailer placement with FEMA following the Camp Fire.

[4] Observed as a member of the Camp Fire Hazardous Tree Removal Operations IMT.

[5] Fellowship interview with Clay Kerchof, CA Department of Housing and Community Development, 2024.

[6] Fellowship interview with Clay Kerchof, CA Department of Housing and Community Development, 2024.

In floods, homeowners are often dealing with 1–4 ft of flooding, and most of the losses are damages that can be repaired. Homeowners may not have flood insurance because it is not pre-packaged into their policies like fire insurance. FEMA calculates unmet needs by identifying the gap between those needs and what insurance can provide. Without the presence of insurance, such as many flood scenarios, subsidies are greater. The better insured a disaster-stricken community is, the less likely it is to receive federal assistance for recovery if the disaster qualifies.

By contrast, as Clay described, the per unit replacement cost in wildfire is much higher by a multiplier of $134,000.[7] A higher percentage of people have fire insurance because it is attached to their homeowner's insurance policy. If their insurance is considered sufficient, they are ineligible from accessing recovery funding.

The presence of insurance in the case of wildfire explains why long-term recovery allocations are so small relative to unmet needs. Clay explained, "insurability is the invisible hand," driving housing recovery, because of how unmet need allocations are determined.[8]

If flood recovery is generally more efficient and better resourced, can we assume flood-impacted communities recover faster and more fully than wildfire-impacted communities? It seems we won't know the answer until the industry is fully designed, which likely means we have generations of wildfire recovery trial and error in front of us.

6.1   Recovering in Reverse

Marci Ryther, a mortgage lender with a local bank, agreed to speak with me for the fellowship. I wanted to understand her experience interacting with clients impacted by the Camp Fire. In the months and years after the fire, Marci met with clients who could afford to rebuild and clients who could not. Without revealing their identities, she walked me through their motivations, decision-making, and the consequences of each decision. What she explained is relatively straightforward but took me a few examples to understand.

Right after the fire, her office was filled with clients ready to do whatever it took to get back on their feet. They came to her looking for numbers and she had numbers. Similarly, a local architect told me the Camp Fire had been burning for only one day when he was retained to draft house plans. By contrast, on day two of the fire, my stepdad was shaving in a mall parking lot in Sacramento, unsure if his house had burned.

Marci admitted to doing hard things like talking people out of rebuilding or buying too soon, before they had proper resolution on their insurance claims and while the

---

[7] Fellowship interview with Clay Kerchof, CA Department of Housing and Community Development, 2024.

[8] Fellowship interview with Clay Kerchof, CA Department of Housing and Community Development, 2024.

market was still inflated from the housing stock loss. She helped them understand their risk, the market, the consequences of all options in front of them. Many clients went forward without heeding her guidance. They invested every penny they could scrape together to buy or build a home that wasn't worth the value of their money. She explained that banks loan on value even though the cost of construction is relatively the same everywhere. She described how people who are rebuilding even years after the fire have to post more cash reserves to build their homes in Paradise because the values don't support the financing.[9]

For fire survivors who purchased homes down the hill in Chico, she describes:

> "Disaster occurred, we had significant housing stock loss which inflated housing prices, people bought out of desperation then wouldn't see that value come back out. They were immediately upside down."[10]

Marci described clients who gave their whole life savings to contractors who poured their foundations then left the area with the rest of their cash. Unfortunately, this was not a one-off situation and she referenced three cases with three different contractors.[11] When I arrived to work at the Town, I was naively unprepared for criminal activity during recovery. The Contractors State License Board still routinely visits our area to check licenses and make arrests.

Marci also explained that the rising cost of insurance reduces buying power. Having to set aside funds to cover the high cost of monthly insurance payments cuts down on the availability of funds to support a mortgage payment.[12]

To better understand the complexity of individual household complications, I spoke to disaster case managers about housing recovery.[13] Disaster case managers work with individual fire survivors to identify unmet needs and match them with available resources. Their work is paid for by non-governmental organizations sometimes funded by FEMA and/or charitable funds. Individual recovery is non-linear and it is often one step forward and two steps back even with case management.

I asked the disaster case manager to provide examples of circumstances commonly found in case management, changing any personally identifying information. What I found were unique examples of similar issues across all of Butte County's wildfires[14]:

1. One client has been in disaster case management on and off since the Camp Fire occurred. They received money from the Fire Victim Trust but became their own barrier by pushing each case manager away. Previous case notes signified this client was difficult to work with so each agency continued to drop them. Once in stable case management, the case manager recognized the client had been deeply affected by the fire and was experiencing the effects of severe trauma. This has been the biggest factor in the home not being rebuilt.

---

[9] Fellowship interview with Marci Ryther, Golden Valley Bank, 2024.

[10] Fellowship interview with Marci Ryther, Golden Valley Bank, 2024.

[11] Fellowship interview with Marci Ryther, Golden Valley Bank, 2024.

[12] Fellowship interview with Marci Ryther, Golden Valley Bank, 2024.

[13] Fellowship interview with Disaster Case Managers, Camp Fire and North Complex Fire, 2024.

[14] Summarized from emails sent by Disaster Case Managers interviewed for the fellowship, 2024.

2. A couple resided in an outbuilding prior to the Camp Fire, in hopes to reconstruct and renovate a house. However, when the fire struck, they found themselves displaced for a lengthy period of time. Upon receiving some assistance from FEMA, they acquired a run-down travel trailer and placed it on their property. During this time, their sole income was from Social Security, with the spouse battling for disability benefits. Faced with financial strain, they reluctantly sought assistance, borrowing money from relatives and friends to cover their daily expenses. As settlement funds from the Fire Victim Trust began to trickle in, they prioritized repaying their debts to those who had helped them initially. Despite being under case management for several years, they remain without a viable housing solution.

3. A client who was a previous renter purchased property after the Camp Fire. They have been living in a trailer on the property for a few years now, in hopes to build a house. They took it upon themselves to tackle all pre-permitting steps on their own such as getting the property surveyed, a septic system installed, a brand-new leach field, and water and power hooked up. They used the majority of their Fire Victim Trust settlement funds on the property. They joined case management after they completed these steps in hopes their case manager could assist with their rebuild. Due to lack of funds, other options have been presented to them such as moving into a local trailer park for cheaper affordable housing. Case managers have observed many pre-fire renters who purchased property after the fire in similar situations. They used all settlement funds on the property, and do not have any funds to put towards a full rebuild.

4. A client visited by code enforcement was in need of help with recovery and was placed with a case manager. He was an owner prior to the Camp Fire and lived alone. He is limited on income as he receives disability benefits. Since the beginning of case management, he began having trouble with trusting the program with his personal information. Once trust was formed and goal planning was established a second obstacle came up when applying for a home loan through the Town of Paradise, he was not willing to give the Town his information they asked for to complete various documents, which made moving forward difficult. He realized that these programs are legitimate and critical for survivors who need assistance with rebuilding and has decided to move forward with applying.

5. A client is a homeowner, who paid for some of the pre-permitting steps after the Camp Fire. The client has been living in a travel trailer since the fire. In hopes to rebuild, other financial strains began to take place. The client lost his job, and is currently going through a divorce. He has used his Fire Victim Trust settlement funds on day to day living. At this time, he is struggling to find work, and unfortunately, does not have the funds and income to rebuild or become accepted into an apartment at this time.

The disaster case manager explained that while they are assisting with the hardest cases five years after the Camp Fire, they do have successes. The month prior to our interview, case managers secured over $46,000 for unmet needs, closing the gap on a client's rebuild.

She explained:

"The funding will assist the client and his family in completing their rebuild. The clients have been working on rebuilding their home since October 2019, while living in travel trailers on their property. These funds will assist with drywall, insulation, flooring and remaining finish items for their home to be habitable. They will finally be able to move forward and recover as a family and are so grateful for the assistance to do so."[15]

6.2 Invisible Disaster

There comes a point after wildfire that the telltale signs of disaster start to disappear. On long drives through Northern California, I can spot burn scars that are ten years old on the hillsides, but visible fire impacts on communities become less evident a few years after the hard work begins. When I first arrived to work in the Town, every vacant commercial lot dripped with melted business signs. Burned car hulls sat in driveways, road pavement was scabbed and scarred, dead and dying trees were everywhere: standing, shattered, half mast, leaning, one day up and the next day down. Evidence of fire was all one could see.

In May of 2023, I helped organize a tour of the recovery for the Rural County Representatives of California, an association serving the needs of rural counties in California. We drove a bus full of elected Supervisors through the Town and up into the unincorporated community of Magalia. As we drove, we described recovery projects, housing reconstruction, tree removal and defensible space requirements, undergrounding and repaving, multi-family housing development, and risk reduction efforts.

When we arrived in Magalia, an association staffer from Sacramento jumped off of the bus with eyes aglow and said, "if you hadn't told me a fire was there, I never would have known!"[16] It was the first time since the Camp Fire I'd heard the disaster described as invisible, and I had mixed feelings. I certainly don't want people to be exposed to the devastation, but her comment almost wiped away what we'd been through. I'm sure she didn't mean to offend—rather I think she meant to compliment the recovery work—but it was a moment when assuming the presence of trauma would have helped.

A few minutes later I was talking to a Supervisor from Mendocino County about his impressions. The Mendocino Complex Fire occurred just before the Camp Fire in 2018. The Supervisor chatted with me for a few minutes then got very quiet. As tears swam down his face, he said he understood so deeply what we'd gone through as he'd gone through it himself. He shook his head and looked beyond me quite a few times, almost seeing into his own past through our fire-stricken trees. He commented on the condition of the lots and the roads, and the many years he knew we still had

---

[15] Email correspondence from Disaster Case Manager interviewed for the fellowship, 2024.

[16] Observed during the 2023 Butte County Tour for Rural Representatives of California.

ahead of us. He suggested we reach out to his Emergency Manager to exchange tips, then rejoined his peers.[17]

These exchanges demonstrate how wildly and quickly disaster recovery perceptions can swing. In these two cases, one was a fire survivor, one was not. Speaking with the Supervisor from Mendocino County, I felt so seen, even with the disaster invisible to others around us. He allowed me into his trauma which created a bridge for mine. On recovery tours, it's typical to move from one person celebrating to the next person crying. I'd say for survivors living day to day, this emotional swing is common, too.

My children's eyes are attuned to the landscape of fire now as well. My oldest daughter who now lives in San Francisco sees a fog bank over the California coast and is relieved to realize it isn't a smoke plume. We see agricultural burns on drives to and from Sacramento and have to talk about the color and size of each smoke plume, measuring its distance from the road, especially when it's windy. Just yesterday we saw one such burn and talked about how the various colors of the plume indicated what was burning. My youngest daughter automatically said, "if it's black, it's structure."

Prescribed burns in the forested foothills set off emotional alarm bells even when we know they're coming. Last week we circulated a control burn email from the Butte Unit of CAL FIRE across our department to give a courtesy heads up and peace of mind. We never assume anyone knows and do our best to reduce the re-traumatization of visible fire. During burn season, 250-acre controlled burns in the foothills that send brown plumes into the sky never leave our peripheral vision.

My kids look at forests differently than I did while growing up. I loved the quiet security of a tight, dense forest. For them, if there's no visibility through the trees then the forest is overgrown and the risk of fire is greater. When Adela was 9 and we were visiting the Bay Area, she told a man in a bagel shop who'd just commented on the rain, "yeah we need rain, but if the snow pack isn't deep enough then we haven't gotten enough precipitation to make a dent in the drought." Growing up in Butte County, she is far more tuned into the land than I was at that age.

Kids in our community are growing up in an age of Red Flag Warnings and Watches. They know wind, heat, and crackly dryness of nearby fuels will lead to Warnings and Watches. As the former Town Manager in Paradise warned me, we will never look at hot, windy days the same way after the Camp Fire. And we will never rest when there's an unaccounted-for smoke plume in the distance.

I remember practicing earthquake drills in grade school while growing up in the Bay Area. In Chico, our schools run Code Yellow and Code Red drills that cover a myriad of threats far scarier than an earthquake. Our kids are already on edge, and the fear of fire pushes them closer with each hot, windy day.

---

[17] Observed during the 2023 Butte County Tour for Rural Representatives of California.

6.3 Opportunity Cost

It is common sense that jobs are lost as a result of direct disaster impacts on businesses, but it's perhaps less expected for job loss to occur as result of recovery. For large federally declared wildfires, hazardous tree removal projects require contractors to work at a massive scale. They also require federal contracting and labor standards compliance capabilities. Because of this, locally owned and operated tree companies and wood processors are less competitive against national companies with federal contracting experience and the ability to expand and contract their labor force.[18]

Sometimes local companies are subcontracted to provide specific or smaller components of the large-scale recovery projects. In many cases, local operators fall prey to the recovery workforce shuffle as their workers join the "big guys" coming in for the government-funded projects. Larger companies can often offer competitive salary and benefits packages—and training—that may exceed local opportunities. Pre-fire, working for a large company may have been less attainable for the local workforce given distance to work sites, but when those work sites are local, the jobs are far more accessible and attractive.

In the hazard tree removal program, contractors were required to identify an available end-use facility for the waste wood when they bid on the jobs. There are only a handful of facilities in the state of California where waste wood can be taken for disposal. The large companies had to reserve capacity at these sites to be eligible as a bidder on the government-funded projects. The end-use facilities would then hold capacity for the large operators, shutting out local companies entirely, or forcing them to transport their waste wood greater distances at a greater cost to the business.[19] When waste wood flooded the market due to these fire-related projects, its value dropped to nothing. Because of this, I heard from a number of local timber operators that they were unable to make ends meet during recovery.[20]

We are accustomed to log trucks and chip/slash trucks leaving our county in droves. The logs are charred or green depending on the season and reason for removal. In any case, they're being transported to a facility where any merchantable wood will be sorted, processed, and sold, and the rest will be chipped and disposed of. On my commute from Chico to Oroville on winter mornings, blackened logs stacked horizontally are caked in snow, chunky snowballs falling from the trucks and shattering on the side of the highway.

Early in my recovery career, I was proud of these trucks leaving our community. It felt like a sign of hope and hard work to see dead and dying trees on their way out. But over time, as I've learned the potential value of woody biomass, I have come to see this wood as a possible fuel source for renewable energy, and seeing it trucked out feels like watching opportunity drive away.

---

[18] Observations from Local Timber Operators during Camp Fire Hazardous Tree Removal Operations, 2020–2021.

[19] Observed during contracting for Camp Fire Hazardous Tree Removal Operations.

[20] Anecdotes from owners of local timber operations during Camp Fire Hazardous Tree Removal Operations.

If fire decimates a community, and recovery projects like tree removal require transport of semi-valuable material out of town by non-local companies who will sell or process it, then surely there is an opportunity cost to recovery. I told a bio-oil manufacturer that if the feedstock is here, the jobs should be as well.

In many communities rich in natural resources, the economy is dependent upon the export of those resources—lumber, water, rock, coal. The economic stewardship model I learned about at the Rural Voices Coalition of California conference inverts this paradigm and suggests natural resources should be reinvested locally to create economic stability and growth. Add fire and reinvestment becomes even more critical to recovery.

For debris removal, we had no choice but to haul excess fire debris out of town that couldn't fit in the local landfill. Given the volume of Camp Fire debris the local landfill did take, years of valuable "air space" was lost instantly. Decreasing the air space accelerated the reduction in capacity and lifespan of the landfill, requiring additional planning and capacity building.[21] So, there is a point at which not all fire remnants can be kept locally.

But trees might be a valuable component to economic recovery if removed responsibly, brought to a local facility or processed onsite, and turned into lumber to rebuild homes or feedstock to generate energy. I understand this approach is happening in recovery where pre-existing infrastructure supports community-scale lumber milling after fire.[22]

If we prepare for recovery to involve the very material we are removing, then perhaps communities can retain and even add jobs post fire. Ideally, we could create a closed loop of forest management + local sorting + local processing = renewable energy sold on the open market or added to the local grid to stabilize energy availability and pricing. This sounds far preferable to losing everything to fire then losing more to recovery.

6.4 Boomerang Effect

It is well documented that disaster creates population displacement. Staff at Chico State studied a sampling of postal code data to show population dispersal across the United States one year after the Camp Fire.[23] Research continues at the University's

---

[21] Neal Road Recycling and Waste Facility Master Plan, prepared for the Butte County Department of Public Works by SCS Engineers, adopted by the Butte County Board of Supervisors in May of 2024. https://www.buttecounty.net/DocumentCenter/View/14973/SCS-Final-Report---Butte-County---NRRWF-Master-Plan-4-17-24-original-with-all-PDF.

[22] Observed from individuals and agencies working in Dixie Fire recovery.

[23] Hansen, P., Mapping a Displaced Population, Chico State, November 9, 2019. https://today.csuchico.edu/mapping-a-displaced-population/.

GeoPlace Mapping Lab where faculty and staff are looking at risk within those relocations, finding that many fire survivors settle back into high-risk areas.[24]

I've come to think of frequent fire relocations as the heartbreaking boomerang effect. People move out of state after their homes are destroyed, then move back within a few months or years, one spouse happy, the other bereft. Or they move to a larger city in California only to make their way back to the small-town life they miss in Paradise, even if only a shred of that life remains. This certainly isn't the case for all fire survivors, but it is not uncommon for people to leave in search of what they've lost, only to realize home is not so easily found.

After the Camp Fire I noticed many of my friends who'd lost their homes relocated quickly if they had children. Local schools closed for months, even schools outside of the burn scars due to poor air quality. When local schools re-opened some classes were held in closed department stores and airport hangars. As CalMatters reports as of September 2025, kids and parents are still recovering from the fire's disruption and long road to normalcy.[25]

At my youngest daughter's 6th grade Open House, her English teacher broke down in tears recounting her time teaching in a shuttered ACE Hardware store aisle after the Camp Fire. As she described the curtains hung to create a sense of place for the students, her face was anguished. I cried with her and for the kids who suffered prolonged lack of normalcy after losing so much. Ultimately, she explained, she gained tremendous gratitude for her classroom at a junior high in Chico and the opportunity to teach our kids.

Rebuilding frustrations are also a cause of the boomerang effect: cobbling together financial assistance, navigating the permitting process, hiring then firing then hiring a contractor, and experiencing labor and material delays. Any hitch in this path for someone who never intended to build a home if it weren't for losing one can tip the scales toward too much. A friend of ours got fed up with the process and expense of rebuilding and moved his family out of state. Immediately his kids began to suffer and what they'd intended to be a fresh start turned sour. They returned after six months, re-enrolled their kids at Paradise High, and are still working through their reconstruction years after the fire.

6.5 Diagnosis

In October of 2020, while I was working at the Town as Disaster Recovery Director, an Electro-myocardiogram (EMG) and blood test confirmed I have a genetic, degenerative neuromuscular disease. The disease runs in my dad's family and is present in every generation. Two days after my dad died in October of 2016, I was rear-ended in Petaluma and spent the next four years looking for answers to explain

---

[24] Murphy, S., Mapping Lab Takes Closer Look at Population Displaced by Camp Fire, November 5, 2020. https://today.csuchico.edu/mapping-lab-looks-closer/

[25] Jones et al. (2025).

sciatica, stroke-like symptoms, nerve pain, and muscle weakness. After four years of bouncing between local specialists, a frustrated physician referred me out of the area to a neurosurgical nurse in Sacramento who said, "that doesn't sound right," and sent me to the physical medicine clinic at UC Davis.

The EMG took place in a sunny three-story brick building in Sacramento off the main hospital campus. It was supposed to confirm the presence of some sort of nerve issue causing the sciatica, but after testing everything on my left side where the pain had flared after the accident, the doctor quietly moved to my right side and began testing again. "Does anyone in your family have a neuromuscular disease?" she asked lightly. I said no and on we went.

At one point she asked if I was cold—always—and wrapped me in a thick hospital blanket out of a warmer in the exam room. Once I warmed up, she took some measurements, taped electrodes to my arms and legs, and pulsed my skin with a round metal probe transmitting electrical currents. My hands and feet jumped off the table in a painful but endurable spasm with each pulse. Over the years I've come to call EMGs the lightning test because of the zapping and flailing.

Casually, the doctor asked me again if I had any presence of neuromuscular disease in my family. Nope, I said and on we went, now into our second hour of testing. She seemed surprised I was a runner though I told her running was becoming harder as I got older and I was transitioning to walking. She moved over my body with a large fork-like tine that shook when she flicked it while I reported if I could feel the vibration in my hands and feet.

Two hours of tests later, and her puzzlement at the results, she asked me one more time about family disease. "No," I said, "although I do have an uncle with a floppy foot disease," referring to my dad's half-brother who I didn't know very well. That's when she sat down and told me about CMT. If I hadn't been a runner my whole adult life, she said, I wouldn't be able to run now. My nerves were unraveling and I was losing my muscle.

An MRI showed glowing along my spinal cord indicating distress where my peripheral nervous system connects up with my central nervous system. The signals from my nerves to my muscles are slowing and may eventually stop due to an unspooling myelin sheath. Without communication from my nerves to my muscles, my muscles will gradually cease to receive messages to move, and they will break down. CMT does not impact cognitive function nor shorten life span, but there's no cure.

My doctor said her lab was working on setting up a clinical trial and I'd be a candidate. I said I was interested and, one year into treatment later—after fitting me with splints and medication to reduce nerve pain—I was called in to test for the clinical trial. I was so enthusiastic about the trial I returned all phone calls immediately and drove down to UC Davis several times for full-day testing. Finally, I was accepted into the trial.

Because I was one the first subjects enrolled in the study, the research clinicians tested trial protocol on me. They regularly apologized for the disorganized visits and long waits as they built the program around me. I assured them they had nothing to apologize for, I was thrilled to be in the study. Any inconvenience was worth being part of medical research for a disease so present in my family.

From 2021 through 2023, I visited the clinical lab at UC Davis every few weeks and months for testing. The schedule was rigorous and so were the tests. I was put on a medication that had to be refrigerated and taken twice a day. After each visit, I left with a new batch of medication in three white boxes that fit into a large blue cooler case with a medical card on the outside. The clinical research staff and I laughed about the fact that the medication was described as "banana flavored" but tasted like sunscreen left out in the sun too long. Bitter banana sunscreen juice. Again, I wasn't bothered.

I was a very good research subject. I took all of my medication and filled out the drug diary twice a day, which was probably the most laborious part of the trial apart from testing. I traveled with my medication and put little white boxes with my name on them in the fridge at work. I never missed a dose.

The trial was long and the research assistants who coordinated my visits changed a few times. They met me down at the lab when I arrived for blood draws, took me up to the research lab for the paperwork, did my EKGs and strength tests, then worked around the doctor's clinic schedule to squeeze in the physician tests. They brought me to the patient clinic for parking validation, reimbursed me for the cost of gas to and from the lab, and conducted phone visits with me every two weeks.

One of my favorite research staff had just returned to work after having her first baby when the trial started, and by the time we were wrapping up my final tests she was recounting her son's 2nd birthday party. We didn't see each other's faces until well into the trial when we were able to remove our masks in the hospital.

I finished the 16-month trial and was invited to participate in the open label extension which is the final drug testing phase before FDA approval. If I liked the medication after FDA approval, I could remain on it with a prescription. I stayed on the open label drug for a few months then made the tough decision it wasn't for me. The formula changed throughout the trial and this last dose had adverse effects. Either that, or I'd been on the placebo and this was the real thing. I may never know.

I exited the study with a heavy heart and cried during my last visit about how much I appreciated the research and what it means to my family. More than anything, being a research subject gave me an incredible view into the world of medical research where labs full of dedicated people are searching for cures. Because this particular lab conducted research on pediatric muscular dystrophy, I'd overlap with kids doing their strength and mobility tests. One particular toddler raised his little arms in victory after laboring up three stairs and shouted, "I did it!" while his dad and the nurses clapped.

The fellowship and the clinical trial overlapped by a few weeks. I applied for the fellowship in September of 2023, completed my final study testing at UC Davis in October, then traveled down to the Stanford campus to join the cohort in November.

CMT isn't curable and I'm no longer treating it beyond the splints I wear at night and on bike rides, and the cane I keep in my trunk for long periods of standing. It is in my blood and in the blood of my children whether they are affected or a carrier. I am the keeper of the family medical history now that I have a list of diagnostics and interventions.

Family members on my dad's side call me with questions on occasion, considering being tested themselves. I think deciding whether or not to be tested is a matter of choosing to live with the known or the unknown. I've decided after much thought not to test my children since our form of the disease doesn't seem to progress until adulthood. The decision is theirs and I will support them whatever they choose.

I told a friend when the fellowship started that I was excited to go from research subject to researcher. I was excited to poke around at something other than myself. I also loved the idea that we were the first cohort to go through the program and it would be tested through our participation.

The similarities between the trial and the fellowship are uncanny. The trial illuminated plenty of things about my body I'd taken for granted or misunderstood, and the fellowship is allowing me to dig into my memories and process what's been hard to access under the weight of my daily responsibilities. Both processes are guided and supported which is necessary for stepping into the unknown.

6.6 Common Reactions

Three years into my disaster recovery career, I took my first course on resilience. It was a teaser tagged on to the end of a meeting, taught by an instructor from the Butte College Training Place. He'd helped develop the curriculum for local businesses and organizations moving through trauma after the Camp Fire.[26]

The only word I can think to describe myself as I sat in that course is brittle. I see pictures of myself from early 2023 and I look eerily normal while inside I felt inches from shattering.

The instructor walked us through a series of slides on trauma: the causes, the effects, the other side of it once you make it through. I am deeply certain now that I knew then I was traumatized by the fire, its aftermath, and my recovery work, but at the same time I'm 100% sure if you'd asked me if I was suffering the effects of trauma, I'd have said no.

One slide had 6 bubbles[27]:

1. Thinking
2. Emotional
3. Physical
4. Spiritual
5. Behavior
6. Relationships.

---

[26] Fellowship interview, Former Instructor for Butte College—The Training Place, 2024.

[27] Resilience training materials from the Trauma Resource Institute.

The bubbles contained words describing common reactions to trauma in each category.

The Thinking bubble said paranoid, nightmares, forgetfulness, poor decisions, suicidal. The Emotional bubble said rage, fear, avoidance, depression, anxiety, guilt, grief. The Physical bubble said rapid heart rate, breathing problems, numb, tight muscles, sleep problems, hypervigilance, trembling. The Spiritual bubble said hopelessness, loss of faith, increase in faith, destruction of self, doubt. The Behavior bubble said isolation, tantrums, eating disorders, addictions, self-injury, violent behavior. The Relationships bubble said angry at others, missing work, overly dependent, irritability.

Huh. I blinked at the list a few times then pulled out my phone and took a photo of the slide. This was going to take some studying.

What struck me was finding the seemingly random quirks I'd developed over the past few years collected on the same list. I know anxiety and depression are common reactions to trauma, but here were all the issues I'd randomly mentioned to people over the last few years: breathing problems, hypervigilance, trembling, the on-and-off clutch of my eating disorder.

I'd told my doctor at a recent visit that my throat sometimes felt like it was closing and I worried about my breathing. I told a reiki specialist I'd seen a few times about the hypervigilance and extreme startle reflex, but chalked those up to being tired all the time. I've had an eating disorder since my early twenties and though I consider myself mostly recovered, I've spent enough time in treatment to know the signs of recurrence. Here and there I saw the signs.

I sat in that resilience training and wondered if I'd been fooling myself. Was I a poster child for common reactions to trauma? Not even occasional or rare, but common? Once again, I had overlooked the obvious.

All my life people have called me 'tough' because that's the image I cultivated. On social media I would boast about my ability to run toward my fears instead of running away from them. I would explain this wasn't because of bravery but because I felt 49% terror and 51% conviction, just enough to lean in. This wasn't humility, it was honest self-expression that probably fooled me into thinking I'd be successful in the disaster recovery field despite my doubts. Terrified but determined was my motto. I told a woman on a work tour recently that I chose between my peace of mind and the lessons of a lifetime when I said yes to this career change on a 2% margin.

When Adela was a young teen, I bought her a necklace that said "fearless" because she seems to come by it more naturally than I do. She attended most of the Paradise Town Council meetings with me as a toddler when I ran the Paradise Chamber, one time lifting my dress from behind as I stood at the podium. That one little moment of involuntarily mooning eradicated my fears of public speaking henceforth. She regularly pushes me past my own limits.

Adela also sat in Chico City Council Chambers for seven years when I was CEO of the Chico Chamber, one time saying, "they talk to each other like they don't know we're watching them," during a particularly contentious discussion. She sees the veil of power then peeks behind it. Recently I delivered some of my black work pants to her on the Chico State campus for a presentation she gave in business class. She tells me she believes she can navigate the working world so easily because I've made my professional wardrobe accessible to her, a small building block to success.

I am embarrassed by so many things from those first years of living and working in Butte County, but especially my youthful swagger and the photos and posts I chose to share on social media. I created a distance between my narrative and the actual experience of living my life, all out of a desire to be liked and to belong. This is precisely where I find myself now, still somewhat caught in the middle.

I like to belong. In fact, I'd probably say I've felt a need to belong my whole life in order to believe I'm good and worthy. I like belonging to associations, to friendships, to non-profit culture, to a downtown where everybody knows my name. I like to be thought of as a participant and a contributor so I am trusted enough to belong. For many, many years I worked hard to belong.

The price I've paid for belonging is the absence of worthiness when I don't. The common reactions to trauma are a barrier to belonging which I'm certain has impacted far more people than just me. In fact, I'd put that number in the tens of thousands following the Camp Fire. In the Spring of 2025, Butte County Behavioral Health ran a media campaign proclaiming the suicide rate in Butte County is 59% higher than the State average.[28]

My social system was not directly disrupted by the fire, but it was deeply disrupted by the alienation of the aftermath and recovery. I see this in the professional choices we've made in response to community needs, and how we turn our common reactions to trauma against others who are trying to move on.

Sitting in the first resilience training that day, I saw a stranger looking back at me from that slide, but I also felt seen. The reconciliation of both is what I'm working on now. I've never been a naturally confident person but I've used that 2% to squeak by, and in the past it has been enough. I'd say the inverse is true at the moment and it has resulted in my withdrawal from friendships, my silence on social media, and the tight hold I have on this story. I see no evidence of belonging or of being liked. Instead of pushing past terror into the sliver of conviction that's welcomed me my whole life, I've leaned out as far as I can while staying in the game.

After the training I was so disoriented I got in my car and drove in the wrong direction, crossing several lanes of traffic needlessly. Thankfully, turning around gave me time to cry all the way back to my office. The instructor asked for feedback after the lesson and I was honest. I said it was unfair to crack us open then send us back into our afternoons without a wind-down or a close-out. We didn't get to the resilience part, either, at least I hadn't. He acknowledged that he and the other training staff felt the same way and would handle these mini trainings differently in the future.

------

[28] Data provided by the Butte County Department of Behavioral Health, 2025.

6.7 River Fire

In 2024, I went on a wildfire and biomass tour in Placer County. It was a well-organized event with seamless transportation and great food. Attendees were well cared for except for getting back to our cars much later than expected due to incredibly informative tour guides at the lumber mill in Lincoln. The logistics were nearly invisible and on the surface it was a success.

Emotionally, however, I was jostled by the day, and I wasn't the only one. Our first stop was the River Fire burn scar where flames had reached just inside the outskirts of Colfax, a lovely town in the Sacramento foothills. Colfax looks strikingly like Paradise before the Camp Fire: ranch-style homes tucked onto large lots in mature forest pockets, partially obscured from the road by thick overgrowth, and great views.

Colfax has a population of 2,000 built around a charming, historic downtown. As the roads wind up into the hills, the views of the Sierra Nevada mountains stretch out for miles above the oak-filled valley. Near the top of the ridge are larger homes with sprawling, panoramic views.

The River Fire occurred in 2021 and is thought to be of human origin. It burned approximately 2,600 acres, and destroyed approximately 108 structures, in two communities spread over two counties.[29] The steep ridges leading up to the homes are filled with the kinds of matchstick trees a long, hot fire leaves behind. Behind the damaged forest outside of the burn scar, healthy trees and meadows cover the green foothills. Here, it isn't fire as far as the eye can see like in Butte County; rather, there's an unsettling contrast of before and after.

Shortly after the tour I happened to sit on a national rural recovery training with a woman familiar with the River Fire recovery. She called it a "low interest disaster," meaning it hasn't caught the attention of the media nor the jurisdictions enough for a leader in disaster recovery to step forward.[30]

My impression of the River Fire is that it left both of the communities it touched mostly intact, which didn't lend itself to drastic changes in community lifestyles for either. This is not a critical mass "wake-up call" fire, it is an unfortunate event, which individual homeowners—survivors and those in standing homes—are left to interpret and react to in their own ways. This is not to say the loss of 100 homes and the disruption of those households was not a tragedy, it very clearly was, this is only to say that given the scale of the communities it touched in a mostly suburban setting, the River Fire didn't impact everyone in the same way.

On the tour of the River Fire, a speaker shared his process of treating his 20 acres just over the hill beyond the burn scar to reduce fire risk. It is very expensive for private homeowners to treat their land, and/or it requires grueling physical work. The speaker explained he is doing most of the work himself and it is taking him years to chip away at his acreage. Several homes leading up to the fire line are still shrouded in dense brush and trees. It's hard to tell from the outside if that is by choice

---

[29] Placer County proclaims local emergency due to River Fire, August 6, 2021, County of Placer, https://www.placer.ca.gov/7471/River-Fire-local-emergency.

[30] Observed while attending the Rural.

or by necessity, due to a lack of resources for defensible space management. Many homes just beyond the burn scars in Butte County are in much the same condition.

I had a hard time lifting my head while standing in that burn scar, partially out of the hardship of seeing more wildfire destruction, and partially out of respect for the survivors. I now know from studying the Camp Fire that it's exploitative to stand in a burn scar without paying proper respect to the survivors. Instead, I kept my eyes on the charred chips and slash beneath my feet, the black bark and stumps revealing trees that were felled and chipped right where we stood.

An attendee asked how long it would take the forest of matchstick trees to regenerate. In this stretch where the fire burned hot, there is 100% tree mortality. I heard someone mumble next to me, "10 years?" I think we guess 10 years to assure ourselves the landscape will return to its natural beauty within our lifetimes. The real answer is at least 80 years with the right soil treatment and proper maintenance which, as the speaker explained about his own property, takes constant care, cost, and effort. If he took more than a decade to treat half of his 20 acres, who would take on the work and cost of restoring these 2600? It depends on who the land belongs to and, as the speaker said, we can't tell them what to do.

A slide from the rural recovery training said, "The greatest loss is often devastation to the land itself."[31] I am in mourning when I visit burned lands, almost as if they are burial grounds for nature. In many cases, these lands are part of tribal territories that are burial grounds.

After the tour, the facilitator sent out an evaluation form. I gave the tour high marks then proposed a trauma-informed approach next time that might soften the exposure of wildfire on people who have experienced it themselves, and deepen the experience for newcomers. I said the approach should involve language acknowledging the presence of trauma and the desire of the tour facilitators to reduce re-traumatization.

At one point on the tour, we posed for a group photo and as everyone stood still for the camera someone shouted, "wildfire!" as if to say, "cheese!" This sent an uncomfortable ripple through the crowd and another person countered with, "say trees!".

It is not enough to witness the devastation and to walk away. Wildfire feelings are too big for most people to swallow anymore, whether they've been personally impacted or known and loved someone who has. Just the loss of the land is enough for grief to spill over into public view. By acknowledging the presence of trauma in ourselves and others, we set the tone for safely sharing these experiences, and we allow big feelings to be present even amongst strangers.

---

[31] Training materials from the Rural Domestic Preparedness Consortium—FEMA, 2024.

6.8   Secondary Trauma

After the resilience teaser, I worked with our team to bring the instructors to our office. Together, they designed three classes especially for the emergency management and long-term recovery teams which took place in the Fall of 2023.[32] The resilience training was intermixed with instruction on standard interpersonal relations, which made the training applicable to non-disaster workers, too.

I've tried quite a few self-help and help-me approaches since entering the field of disaster recovery. In five years, I've seen three counselors, a reiki specialist, a certified life coach, and countless doctors for a variety of anxiety-driven ailments. The life coach, my friend Caryn, took the approach of setting up a few phone appointments with me to dig in to the effects of secondary trauma. Before the first call with her, I had a hunch what was holding me back and I tested the topic with my husband. I couldn't even think about Dave Daley's cow story without weeping. It was time to let it out.

The County seat is in Oroville which means I drive from Chico to my office in the morning, and back to Chico in the evening. The drive is striking at all times of day, in all seasons: rolling hills, low fog, high clouds, wildflowers, blooming orchards, bee boxes, and big skies for rainbows and storm watching. One morning a helicopter deftly lifted and dropped into the pink orchards, up and down, over and over, lifting cold air up and pushing warm air onto the trees to keep their delicate blossoms from freezing. It's an expensive endeavor for occasional springtime freezes, but not as costly as losing a whole crop.

During winter months the fields are full of grazing cattle. Migrating birds fly overhead in sharp Vs or long strands of dark rivers in the sky. Calves run and jump playfully like little see-saws, while mature cows keep their heads in the grasses morning, noon, and night. One morning a black cow munched happily on the wrong side of the fence while a CHP car sat between it and the highway until the rancher arrived. When the wildflowers are blooming, thick pools of purple and yellow pop brightly in the fields.

Driving back and forth through these spacious working lands allows me time to tune in to my thoughts and feelings. Watching the cows graze often brings back Dave's story of the pregnant cow standing in the puddle. I think of her as I watch the healthy, robust cows eating under the winter sun while their babies push under their bellies and frolic around them. When the slotted trucks arrive in late spring and the cattle disappear for the summer months, I now know they've gone to the mountains.

On my commute one morning I saw a female deer sitting in the median of the highway between two fast lanes going in opposite directions. That's oddly unsafe, I thought. Her ears perked up straight and her body was still. As I approached her, however, I saw she wasn't sitting. Her legs were splayed broken underneath her, and she was unable to move. Still, she looked ready for whatever was next which surely must be better than this. As I drove by at top speed, I knew there'd be no saving her, these were her final moments. Animal control would arrive to take care of her,

---

[32] Instruction customized for Butte County employees by Butte College—The Training Place, 2023.

much as Dave had taken care of his pregnant cow. For the next week, I wept on my commute and still feel tears when I think of her.

During my first coaching session with Caryn, I told her the cow story over the phone in the parking lot at work, barely choking out the words. She agreed if it wasn't the beating heart of my grief it was pretty close. She gave me my first assignment: see if I felt safe enough with anyone to share the story. Shortly after this coaching call, I was on tour of the North Complex Fire scar with Jim Houtman who talked about CAL FIRE counting wildlife losses.[33]

I wanted to talk about Dave's cow with Jim and my colleague in the car, but I couldn't get the words out without crying, which didn't feel professional, so I looked out the window instead. Still, if I had been able to talk about it, it would've been a safe place to do so which is all we were after. Caryn knew my world had gotten very small and she was challenging me to find spaces and people in my life who could withstand and maybe relate to this level of grief. To this day, this is how I test people and places—can I tell the story here? If the answer is yes, I know I'm in good hands. My hope is to one day have an abundance of stories to tell, and an abundance of people to tell them to, and vice versa.

During the next session Caryn gave me my second assignment. She knows my daughter, Heidi, loves arts and crafts so she told me to make something with my hands to honor the cows. It could be a painting, a poem, a picture, whatever came to mind, and to involve Heidi if that felt right.

On a trip to the local craft store I searched for inspiration. I'm not super crafty so I let myself wander the aisles before finding a row of felt sheets in various colors. The options were limited but there were several shades of green, brown, and yellow. Suddenly I could see it, a scene with happy cows in the hills, grazing on grass, safe and sound in the sun. Heidi wasn't entirely on board with my vision but it was all I could come up with so we bought the felt and went home to craft.

I'm a bit of a perfectionist but I knew the spirit of this project was to let my feelings out, not to control the outcome or try to make something of value to anyone but me. It was the doing that was important, not the end result. I grabbed the scissors and started shaping hills, the sun, and cows out of the felt, sticking them together in layers. Heidi eventually saw the vision and joined in, cutting out cows and helping me organize the scene. To avoid overthinking I started gluing right away, pausing only to think of the scale of the animals on the hillsides and mountains.

At one point, when the felt scene was almost done my husband came over to see what we were doing. "Are those….dogs?" he asked, "or…gorillas, or elephants…is that a baby crawling?" As he pointed to our creatures Heidi and I started laughing hysterically. I tried to shoo him away but fell over weeping with real tears of laughter as he examined our beautiful artwork.

For several minutes I cried and laughed and laughed and cried, wiping my eyes and gasping for breath. It was the loveliest moment imaginable, my cows, my family, my grief, and my ability to laugh at myself, all wrapped up in a single piece of art.

---

[33] Information provided by Jim Houtman, Butte County Fire Safe Council and former fireman for the City of Chico, CA, 2023.

Caryn was a genius for suggesting this. She knew making something was the only way to change the shape of the pain and its grip on my heart, and bring a little bit of laughter back.

Reference

Jones C, Tagami M, Lurye S (2025) Seven years after california's deadliest fire, schools—and kids—are still recovering. CalMatters. https://calmatters.org/education/k-12-education/2025/09/wildfire-california-schools/

Chapter 7
Research

7.1 Quantifying Individual Recovery

Shortly after I discovered the obvious—that there is no single definition of recovery—I began to think about the pros and cons of quantifying individual recovery. As Professor Siembieda says, it is a choice of practitioners not to quantify recovery because of how dynamic and complex the process is. But then how do we measure and report on our progress?

I run into the same problems quantifying recovery as I do defining it: who gets to decide? As recovery instructors will tell you, all disasters are local. As global as COVID was, its impacts were far more local than universal. In Paradise, COVID was a blip, in other communities COVID was a full-blown disaster.

Disaster impacts are precisely quantified: acres burned, number of fatalities, percentage of structure damage and destruction. Disaster response is measured in numbers, too: personnel and equipment assigned, percentage contained. My fellowship mentors encouraged me to think about what quantifying recovery could mean, so I started listening for it in the world of disaster recovery.

At the River Fire tour, a woman from a neighboring county told me she was watching Butte's recovery closely. She said her city recovered 80% of their homes lost in a recent fire within two years. This is a very impressive statistic and I congratulated her and her community on their hard work, but what I wanted to know was:

- How many homes had they lost?
- What was the percentage of total housing stock lost?
- Were structure damages contiguous or scattered?
- Were the destroyed homes hooked up to municipal water and sewer?

That's just the tip of the iceberg in terms of understanding the pace and scale of recovery.

Professor Siembieda encouraged me to focus on viability rather than quantification, but even viability is complex. In an interview with our Assessor's office, I

K. Simmons, *Three Fire Mountains*, https://doi.org/10.1007/978-3-032-17343-0_7

learned they'd done an analysis of taxable base rates for recovery to understand the impact of legislation to extend property tax breaks. In their analysis dated May 23, 2023, of 14,200 total assessments, they found 4,671 properties had been sold, 2,043 properties had been rebuilt, and 1,284 property owners had moved out of state. This left 4,165 property owners who may seek relief which, the Assessor argued, was highly unlikely given the number of people who would not rebuild based upon age, income, ability, and a variety of other factors. She referenced her own in-laws who, in their eighties, lost a home they would never rebuild.[1]

For disaster relief analysis, she explained that land is included if the base is transferred under AB 556, and not included if the home is rebuilt under AB 1500. The calamity base transfer only applies to a similarly valued home. For many people upgrading, this means the calamity base transfer does not apply. For example, a property owner had a 1,000 ft^2 2 bedroom/1 bath home worth \$250,000. If they purchase a 1,600 ft^2 home with 3 bedrooms and 2 bathrooms worth \$400,000, their original base would be reinstated at the time of the transfer, and they will have a secondary new base for the \$150,000 upgrade.[2]

AB 556 states that the property tax base year value of real property that is substantially damaged or destroyed by a disaster, as declared by the Governor, may be transferred to a comparable property located within the same county that is acquired or newly constructed within five years after the disaster as a replacement property. With the Assessor's advocacy on behalf of Butte County, the bill extends the initial five-year time period by three years if the property was substantially damaged or destroyed by the Camp Fire.

AB 1500 authorizes the owner of property substantially damaged or destroyed by a disaster, as declared by the Governor, to apply the base year value of that property to replacement property reconstructed on the same site of the damaged or destroyed property within five years after the disaster if the reconstructed property is comparable to the substantially damaged or destroyed property. The bill extends the five-year time period by three years if the property was substantially damaged or destroyed by the 2018 Woolsey Fire or by the 2018 Camp Fire.

In both legislative cases, data was used to analyze the pace at which recovery is occurring, and the circumstances surrounding properties that are not being rebuilt, i.e., if they are sold or the owner relocates out of state. This analysis showed that not all properties eligible for disaster tax relief would seek this relief or, over time, be eligible for it.

Quantifying recovery means we can see just from this data that the number of lots sold is more than double the number of properties rebuilt.[3] The Assessor's staff and I talked about the value of looking at relocation data to determine the likelihood

---

[1] Data produced and analyzed by the Butte County Assessor's Office in 2023, and provided in a fellowship interview, 2024.

[2] Data produced and analyzed by the Butte County Assessor's Office in 2023, and provided in a fellowship interview, 2024.

[3] Data produced and analyzed by the Butte County Assessor's Office in 2023, and provided in a fellowship interview, 2024.

of a rebuild, perhaps tracking this number from the time of the incident for the next several years. It's quite possible there is a correlation between an owner's physical distance from their property and their ability or interest in rebuilding, and tracking this information annually by a local government can provide insight into the rate at which the tax base might be restored.[4]

CAL FIRE produces a Damage Inspection Report (DINS) for major incidents. The DINS process organizes structures into five types—single family, multi family, mixed use, commercial, other/minor—with four categories of damage:

1. Affected (1–9%)
2. Minor (10–25%)
3. Major (26–50%).
4. Destroyed (> 50%).

Since local government has access to DINS information, a City or County could potentially use the parcel data on file with the local Assessor (wells, septic, owner-occupied), layer in the statistics gathered by the State on available resources (FEMA individual assistance, presence of insurance, disaster loans disbursed, unmet need calculations), factor in the availability and readiness of settlement payments, then tap economists to advise on market conditions. Taken together, this data could help determine the rebuild feasibility of single-family residences in categories 2–4. Category 1 could be habitable with remediation and likely not require the full resources of a rebuild.

Here are some examples of Damage Inspection Reports:

Dixie 2021 DINS Public View.[5] https://gis.data.cnra.ca.gov/datasets/CALFIRE-Forestry::dixie2021-dins-public-view/explore.

Camp Incident Damage Inspection Report[6] https://www.nist.gov/system/files/doc uments/2020/11/16/2018 Camp Incident DINS Final Report.pdf.

Figure 7.1 is from the Camp Fire DINS report.[7] If we're looking at the feasibility of Camp Fire recovery overall—specific to individuals—we might apply the parcel metrics to the top row from minor through destroyed. In the case of this disaster, that's still the vast majority of the damage (fellowship interviewees indicate flood impacts trend higher on affected versus destroyed, the opposite of wildfire) but weeds out what is likely not residential and/or subsidy-dependent to get closer to a rebuild estimate. Then, we might look at interventions like incentives (State-operated loans

---

[4] Observation from the Butte County Assessor, fellowship interview, 2024.

[5] Dixie 2021 DINS Public View produced by CAL Fire, hosted by the CA Natural Resources Agency GIS. https://gis.data.cnra.ca.gov/datasets/CALFIRE-Forestry::dixie2021-dins-public-view/explore.

[6] Camp Incident Damage Inspection Report, CABTU 016737, November 26, 2018, Submitted by Nick Wallingford, Damage Inspection Manager. https://www.nist.gov/system/files/documents/2020/11/16/2018 Camp Incident DINS Final Report.pdf.

[7] Camp Incident Damage Inspection Report, CABTU 016737, November 26, 2018, Submitted by Nick Wallingford, Damage Inspection Manager. https://www.nist.gov/system/files/documents/2020/11/16/2018 Camp Incident DINS Final Report.pdf.

Overall (all jurisdictions)

Category of Damage	Affected (1-9%)	Minor (10-25%)	Major (26-50%)	Destroyed (>50%)	Grand Total
Single Residence	412	47	3	13696	14158
Multiple Residence	21	3	1	276	301
Mixed Commercial/Residential	1	1	0	11	13
Non-residential Commercial Property	76	18	8	528	630
"Other" Minor Structures	89	32	15	4293	4429
Total	599	101	27	18804	19531

Fig. 7.1 Chart from the Camp Fire DINS Report showing loss of structures across all jurisdictions

and grants) or barrier removals (programs to offset septic repair, for example) to measurably increase that estimate to set a goal for the first five years of recovery.

Professor Siembieda explains there are two approaches to individual recovery: incentivize or remove barriers.[8] In cases of rural wildfire recovery, removal of barriers may be a more affordable route if grant funds are available to offset fees and lower the overall cost of reconstruction.

Using this method, imagine an urban fire with high-density, owner-occupied housing connected to municipal services, and a minimum household income of 120% of the Area Median Income (AMI). Imagine 30% of the total housing stock is impacted by fire, sustaining damages from 10% to greater than 50%. Then, let's say using DINS plus parcel metrics the rebuild estimate is 70%. Imagine removing barriers and incentivizing master planned developments through permit streamlining driving the rebuild estimate up to 80%. If 70% of your housing stock is untouched and you can recover 80% of what you lost, your chances of community viability five years post disaster are very high.

Now imagine a rural wildland fire damaging or destroying 5,000 homes in the wildland urban interface which represent 80% of the total housing stock. The rebuild feasibility comes out to 30% and barrier removal is too expensive for the local government. With 20% of their housing stock remaining and only 30% of what they lost coming back, community viability is low.

I say all of this knowing full well it is not that simple; recovery cannot be worked out on a piece of paper. I also feel very triggered as I write this—similar to how I felt when the researchers suggested we only rebuild in the center of town. A mathematical approach to recovery is off-putting to me. It is clinical and sterile to reduce personal choices to statistical likelihoods after a tragedy of this size. But, is it one way to understand individual recovery, or a way to supplement what's clear from working directly with survivors? Yes, I believe so.

---

[8] Communication from William Siembieda, Professor of Planning, CalPoly, over the course of the fellowship, 2024.

7.2 Community Recovery Planning

The rural recovery training I attended referenced the Joplin, Missouri, EF5 tornado recovery as a model for post-disaster community planning.[9] The tornado was the nation's deadliest and costliest tornado on record and within 6 months the recovery team had a community-led recovery plan in hand that, as the instructor emphasized, received a standing ovation when it was released to the public.

A community working together to wrap expectations and a timeline around recovery makes a lot of sense. A plan gives a community a sense of comfort, directional unification, and shared expectations. If an outside agency working independently of local government, organizations, and residents came in and handed out a recovery plan, the community would likely revolt.

In 2019, a consultant led Paradise residents through the development of the Long-Term Community Recovery Plan following the Camp Fire.[10] This plan identified community projects and priorities to guide the Town, local agencies, organizations, and the community at large toward a shared vision of recovery and resilience. When I worked at the Town I regarded the plan as the highest form of recovery direction, led as it was by the people the fire had impacted. The consultant worked with the Town to update the plan in 2022.[11]

An attendee on the rural recovery training shared a downside to planning recovery I hadn't considered. He referenced a recovery plan that announced a date by which the local school district would reopen following disaster. For many parents longing to re-establish routine and familiarity in their kids' lives, this news was a great relief and something they planned their lives around. For others, it added pressure to a commitment they weren't ready to make. It represented a ticking countdown to a significant marker of recovery that accelerated their decision to remain in place—a decision they weren't sure was right.

In the end, recovery can only be defined and measured by each person experiencing it. To measure collective recovery in a perfect world, a plan must be set in place defining specific outcomes according to a set timeline, with established roles and responsibilities to maintain accountability. The plan must be fully funded for implementation and reviewed occasionally for updates and adjustments as the reality of recovery sets in.

The rural recovery instructor shared a list of 4 key indicators that a community has recovered successfully[12]:

---

[9] Rural Domestic Preparedness Consortium—FEMA, 2024.

[10] Long-Term Community Recovery Plan, Paradise, CA, June 2019, Urban Design Associates, available on the Town of Paradise web site. https://www.townofparadise.com/sites/default/files/fil eattachments/recovery/page/42792/062519_final_recovery_plan_compressed.pdf.

[11] Long-Term Recovery Plan Update, Progress and New Priorities, Paradise, CA, November 2022, Urban Design Associates, available on the Town of Paradise web site. https://www.townofparadise. com/sites/default/files/fileattachments/recovery/page/42792/ltrp_update_11.1.22is.pdf.

[12] Rural Domestic Preparedness Consortium—FEMA, 2024.

1. Meets its priorities to overcome disaster impact.
2. Reestablishes an economic and social base.
3. Instills confidence in local citizens and businesses.
4. Rebuilds community in a more resilient way.

This can be used as a guide but is still subjective. Should recovery ever be declared successful if even a single household is left permanently displaced or unhoused by disaster? In my experience, recurrent wildfire recovery is like filling an ocean one raindrop at a time; as the ocean fills it evaporates, starting the cycle over again.

7.3 Triaging Community Recovery

If setting a goal for housing recovery then working backward isn't the right approach, I imagine recovery professionals running through a decision tree full of questions that may be applicable to community recovery. The decision tree would act as a triage step following response and during early recovery, to begin to assess the overall viability of community recovery. Much like Professor Siembieda's recommendation, the word viability appears over and over in the limited long-term recovery research I can find.

Merriam Webster defines viability as the ability to succeed or be sustained. After disaster, are we working toward viability or are we working toward recovery? The down payment approach certainly leans toward viability (Fig. 7.2).

Each fire has a unique cause which has implications during recovery. Wildfire is a catchall word for a disparate series of causes and effects. No two fires are the same. At the County, we say to each other, "you've been through one fire, you've been through one fire," meaning the skills may be transferrable but they won't be a perfect match because needs and variables differ.

I describe working on recovery from two different wildfires as a "bifocal" experience. Each fire requires a different lens with a different depth and strength. I can see both fires within the same frame but I can't focus on them at the same time because the overall picture of recovery blurs. Now that we've layered in even more fires and recovery processes, we might as well be looking through a kaleidescope.

The Camp Fire was ignited by a faulty electrical transmission line owned and operated by PG&E.[13] Lightning is the widely reported cause of the North Complex Fire, originally called the Bear Fire before it merged with others.[14] Camp Fire recovery involves resources from settlements between PG&E and impacted jurisdictions and survivors. North Complex Fire survivors have no such settlements. This single difference in resources defines much of the recovery pace and scale.

---

[13] Indictment filed with the Superior Court of the State of California for the County of Butte, on March 17, 2020. https://www.buttecounty.net/DocumentCenter/View/1882/Camp-Fire-Indictmen t-PDF.

[14] Mapping of the North Complex Fire, UCMapping 4, July 10, 2022. https://storymaps.arcgis.com/ stories/e905b0aa9d224c95ac8fa7b579cbb66e.

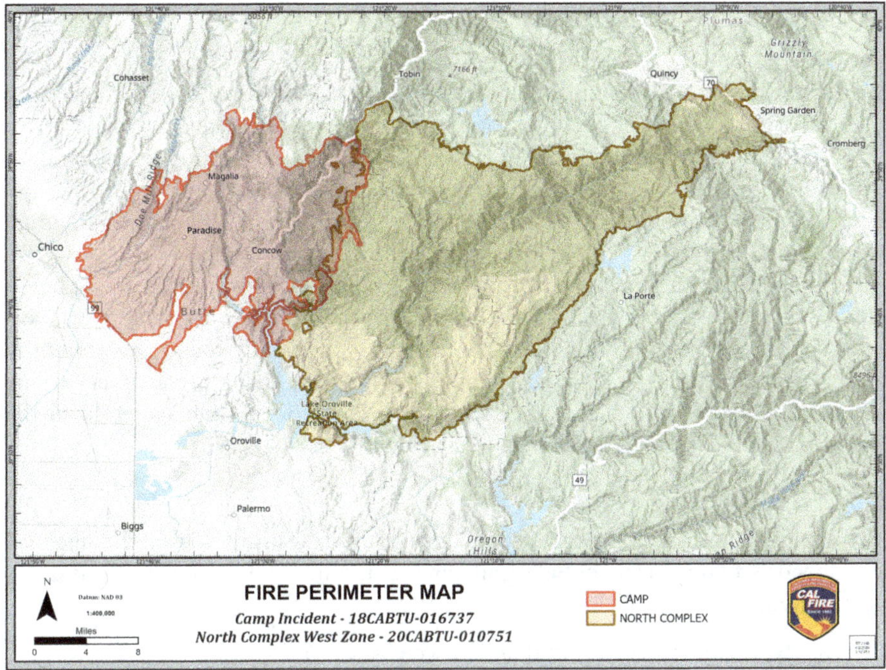

Fig. 7.2 Camp Fire and North Complex burn scar map produced by CAL FIRE

Both fires significantly destroyed entire communities and large wooded areas. Much of the Camp Fire destruction, however, occurred in a municipality with localized staffing, services, and capacity. All of the North Complex Fire occurred in unincorporated areas of Butte County where small communities lack dedicated and defined resources, served as they are by a workforce operating throughout the county. Again, just this distinction in jurisdiction impacts the ability and capacity for recovery.

If I'd never entered the field of disaster recovery, I might think recovery outcomes were entirely dependent upon the number of acres burned and fire duration, as these are the statistics we hear about in the news. From a firefighting perspective, these details are critical for response. But for recovery, they are less consequential than the cause of the fire and who is responsible for rebuilding the community. As the recovery professional explained about the River Fire, two different cities in two different counties lost homes and, in her opinion, no one really stepped forward to take charge. Lack of leadership may have a far greater outcome on recovery than how long a fire burned.

7.4 Recovery Triage Tool

As my fellowship research unfolded, and a simple mathematical equation on paper didn't feel right for assessing recovery or community viability, I began to think about the broad categories we need to understand as disaster recovery professionals. Rather than define and quantify recovery in finite terms, I started to consider questions that might help newcomers to disaster dig deeper into the various tenants of individual and community recovery.

I developed a list of somewhat overlapping questions in various categories to begin to reveal the recovery picture. I worked with Ashima Tshering at Stanford University to design a triage tool to indicate categories that typically relate to community recovery and categories that inform individual recovery. Using the recovery triage tool, professionals might begin to assess the severity of the damage, fold in the aftermath effects, and use both to frame out possible next steps.

Assess the Aftermath

After quantifying disaster impacts such as structure and utility loss and damage, I recommend assessing the potential significance of the aftermath using these questions to get started:

Q. What type of disaster occurred and in what season?

- Are there concerns about debris flow or flooding in the same or different jurisdiction(s) as the fire impacts?

    - What is the timeline for the concerns, i.e. increased flood risk for the first 3–5 years following fire?

- Will immediate or seasonal weather increase the hazards in the impacted area over the duration of debris and tree removal?
- What agencies are measuring risk and are they communicating?
- Are the people showing up to respond also survivors?

Depending on the size of a wildfire in California, the CAL FIRE Incident Management Team may begin organizing resources to determine the risk of post-fire hazards. Using their data to answer these questions may help anticipate the compounding effects of weather and secondary disaster on response and recovery. A compounding effect may be fewer dedicated resources if the aftermath grows into a larger disaster and the workforce is displaced or spread across a greater number of needs. The aftermath will also be shaped by how many survivors are part of response. Tracking trauma from day one will be critical to providing and receiving adequate support and resources.

Anticipating or adding aftermath impacts to disaster impacts on people and the landscape, within a reasonable timeframe, allows for a more holistic view of what the community will reckon with long term. When the worst has occurred, it's hard to imagine things might get worse, but it's best to understand and accept that possibility going in to recovery.

Fig. 7.3 Chart produced by the California Department of Conservation showing procedural steps of producing a Watershed Emergency Response Team Report

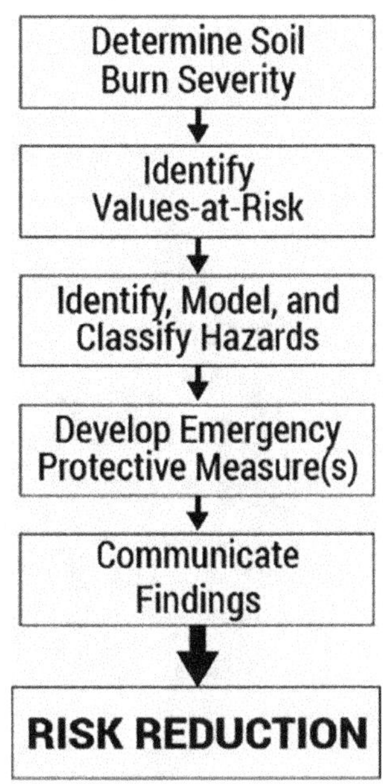

CAL FIRE produces a Watershed Emergency Response Team (WERT) report following certain wildfires that offers rapid analysis and recommendations for mitigating and managing flood and debris flow following fire. A WERT can help disaster-impacted jurisdictions and surrounding communities prepare for aftermath impacts.[15]

According to the California Department of Conservation, WERTs were developed because people and property are at risk of rockfall, flooding, debris flow, and other environmental hazards after wildfire. Their California Geological Survey team co-leads the effort with CAL FIRE who is unable produce a WERT for every fire due to limited resources.[16]

Looking at the larger recovery triage tool, it is difficult to say in which order each section should be analyzed, rather, the suggestion is to ensure each section is considered. Recovery is an expensive and lengthy endeavor, and most questions are geared toward illuminating the immediacy and quantity of available resources to meet demands. Be aware some impacts will be less visible than others from the outset, requiring deeper testing and more time to measure and assess. In Butte County, fire impacts continue to roll out in waves well after disaster, making it nearly impossible to know the magnitude right away. Continue asking these questions through recovery, acknowledging answers may change over time (Fig. 7.3).

---

[15] California Department of Conservation, "What's a WERT?". https://conservation.ca.gov/cgs/bwg/wert.

[16] California Department of Conservation, "What's a WERT?". https://conservation.ca.gov/cgs/bwg/wert.

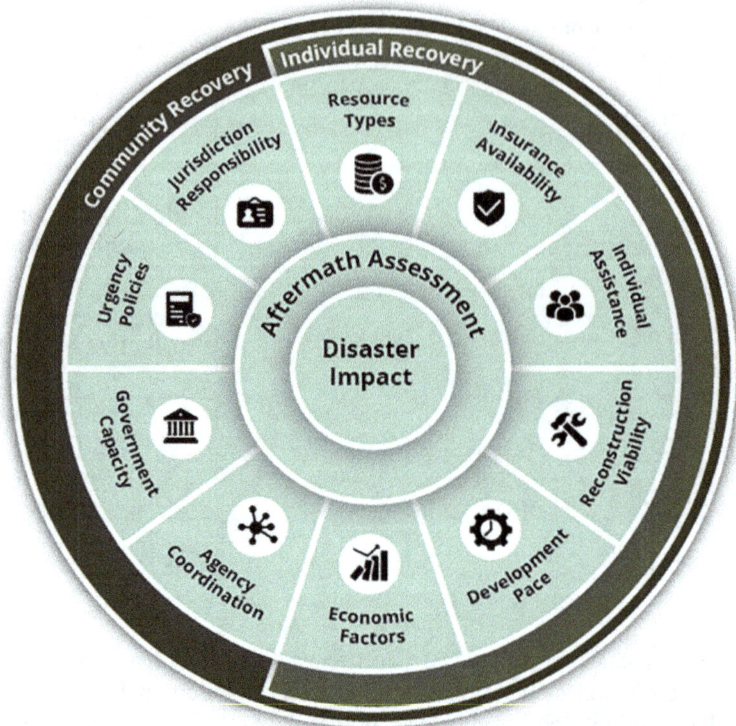

Fig. 7.4 Triage Tool showing various categories of focus for community and individual recovery. Produced by Ashima Tshering with Stanford University at the direction of the author

Recovery Triage Tool

See Fig. 7.4.

Jurisdiction Responsibility

Who is in charge of recovery?

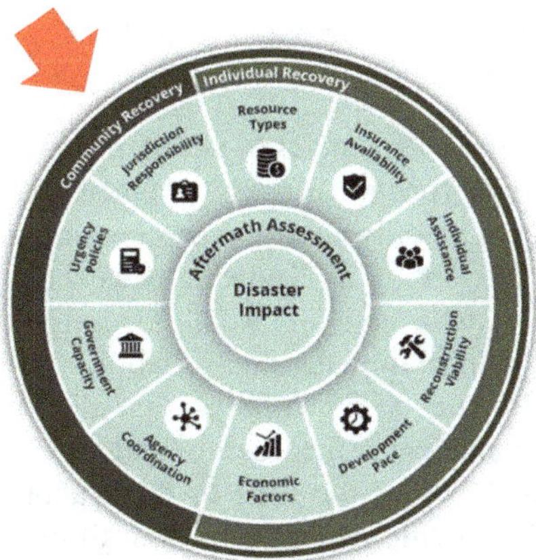

As the disaster impacts and aftermath risks are assessed, it's important to start framing up responsibility for recovery. Understand how many jurisdictions were impacted and how the impacts across jurisdictions are the same/different:

- Are State and/or federally owned lands damaged or destroyed, and how are the responsible agencies responding?
- Are the areas impacted known to be economically disadvantaged and/or high on the social vulnerability index: https://svi.cdc.gov/map/
- Does the impacted area have a high degree of wildfire recurrence?
- Is the population largely experienced in response and recovery?
- Are the majority of homes dependent upon municipal services like water and wastewater, small community systems that are privately owned, or individual wells and septic?

  - Are any water or wastewater systems damaged or destroyed, what is the funding availability and timeline for repair, and who or what agency/district is eligible for that funding?

- What is the status of public and privately owned utilities?

  - Are the private utility companies planning upgrades during restoration like undergrounding or other hardening activities that may need additional time, permits, public messaging, traffic closures, and funds to complete?

- Are the majority of homes located on/around private roads the homeowners will be responsible for repairing? Are the homes generally accessible for recovery and reconstruction vehicles and/or behind locked gates?
- Are any State highways or county-maintained roads damaged or destroyed by the fire, response, or recovery?
- Which communities will be most impacted by debris/tree removal truck traffic, depending on where the end use facility is located?
- Is there a known community gathering spot or key organization serving the burn scar who will step up as a community leader?

- What are the immediate and lasting population changes within the burn scar and all surrounding jurisdictions?

Acknowledging the number of agencies involved in recovery and the scope of their responsibilities may help everyone fall quickly into their lanes. With the exception of permits, private companies generally neither request permission to perform repairs or upgrades nor inform local government. Knowing what is happening within recovery requires paying attention to what's visible from the roadside and on the news.

These questions may help illuminate the availability of funding and immediacy of needed infrastructure repairs within different jurisdictions, and the dependency of housing recovery on municipal infrastructure and services versus self-owned or private systems and roads. This is the time to think about damages across unincorporated areas as well as within cities, and the different service levels within both.

Rural fires and urban fires may require different approaches from the agencies responsible for recovery. Whereas urban agencies may direct recovery funding toward concentrated infrastructure and direct services for repair and reconstruction, rural agencies may contend with disparate population impacts and significant infrastructure damages between them. Small fires impacting dense populations may reach the value of economic losses which trigger State and/or federal funding, whereas, large fires in rural areas where more land but fewer homes are damaged may not.

Answers to these questions may also help determine how prepared and responsive the community and local government will be to recovery, based upon prior fire experience. Overall, these questions may start to shine a light on temporary housing needs and conditions, and the volume of public and private resources needed depending upon how responsibility for land and infrastructure is divided up.

Insurance Availability

Will insurance proceeds be enough for most survivors to rebuild?

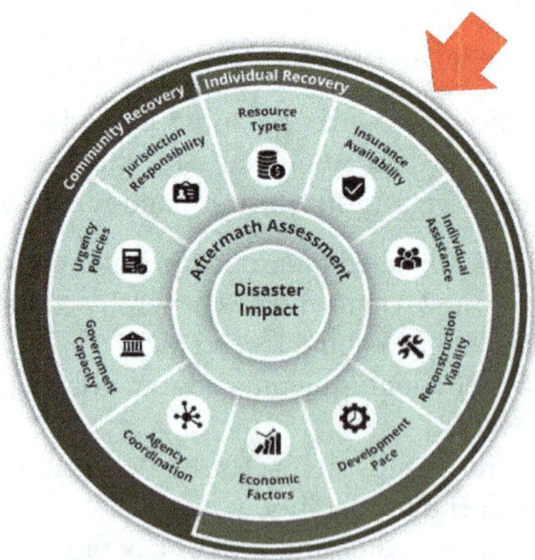

As Clay Kerchof with HCD says, private insurance is the quickest path to recovery.[17] Consider what percentage of the population is adequately insured:

- If households have insurance but it is inadequate for reconstruction, what are the gaps and potential resources to fill them?
- If households lack insurance, is this due to affordability/availability, because their house was paid off and it's not required, or something else?
- What percentage of the population is on the FAIR plan?
- Did the fire occur in an area considered "vulnerable" by the Department of Insurance?
- What percentage of the population received a notice of non-renewal within the last 24 months?
- What percentage are owner-occupants, renters, or second home owners, and what are the varying rates of insurability between them?
- Are new builds insurable in the burn scar?
- Has the Department of Insurance issued any mandates or requirements to insurers related to the disaster?
- Are insurance consumer groups engaging in advocacy on behalf of impacted households?

Understanding insurability and types of home occupancy will begin to frame out the resources available for individual recovery. Knowing early on who has access to those resources, and what and where the funding gaps might be are good indicators of the length of recovery. In California, we currently we have the added pressure of insurance market uncertainty in fire-prone and recovering areas. Consider the status of the insurance market in disaster-impacted areas. Consumer insurance advocacy groups like United Policyholders are ongoing resources for this analysis.[18]

Economic Factors

Is the local economy sustainable during recovery?

---

[17] Fellowship interview with Clay Kerchof, CA Department of Housing and Community Development, 2024.

[18] United Policyholders web site. https://uphelp.org/.

To begin assessing the economic impacts of a fire, consider direct business and workforce impacts:

- How many businesses are damaged or destroyed compared to the total number of pre-existing businesses, and what % job loss does this represent to the total market?
- What percentage of impacted businesses are adequately insured, and/or are receiving subsidies to re-open?
- What percentage of jobs lost are local-serving and dependent upon the pre-fire population?
- What percentage of businesses are relocating nearby with plans to return?
- What is the temporary or permanent shift in industry groups, i.e., are most dentists relocating, are ancillary industries serving the largest impacted employers relocating?
- What are the workforce impacts of population dispersal to nearby towns and to non-commutable distances?
- Are commercial or industrial spaces functional for alternative recovery uses?
- Are commercial and industrial property owners amendable to removing debris and other hazards to prepare their land for temporary uses?

If a majority of local businesses are not lost or interrupted by disaster, they could be a critical linchpin to keeping the economy going during recovery. These businesses may choose to operate in service of response and recovery, adapting their products and services to changing demands.

If a majority of businesses are lost to disaster and are dependent upon population restoration, understanding the decisions of each may help map out who will return, who will relocate and where, and which businesses may permanently close. If jobs are primarily serving local households as opposed to visitors, businesses may need housing to recover before reinvesting.

Of course, some business owners and operators are drawn to recovery and want to be part of community resurgence, and so may choose to rebuild or reopen before the market is ready. While this is risky for the business, it also begins to rebuild trust and interest in the economy, and other businesses may follow suit.

Expect a significant churn in the early years of recovery as businesses gamble on a recovering population and, for reasons inside and outside of their control—like larger economic conditions—may not be in a financial position to sustain operations. Recovery is not invulnerable to additional economic interruptions on a national or international level. Seeing businesses re-open then close after disaster may trigger additional waves of grief. If a wildfire is catastrophic and results in a destroyed community, the economy may not stabilize for several years even with direct business assistance. It's important to expect and normalize this churn over time.

It may be helpful to inventory surviving and/or operational commercial spaces and lots for repurposing for recovery needs. Large destroyed lots, once cleaned, can operate as utility or corps yards for large-scale contractors moving in for recovery. Standing commercial buildings can be used as headquarters for contractors establishing a local presence.

Ultimately, uses of these spaces are dependent upon demand, what the local government allows, and what owners wish to do with their properties. Public–private agreements can be attractive temporary uses of land during the early stages of recovery, creating cash flow when there might otherwise be none.[19]

Reconstruction Viability

Are labor and materials available and affordable?

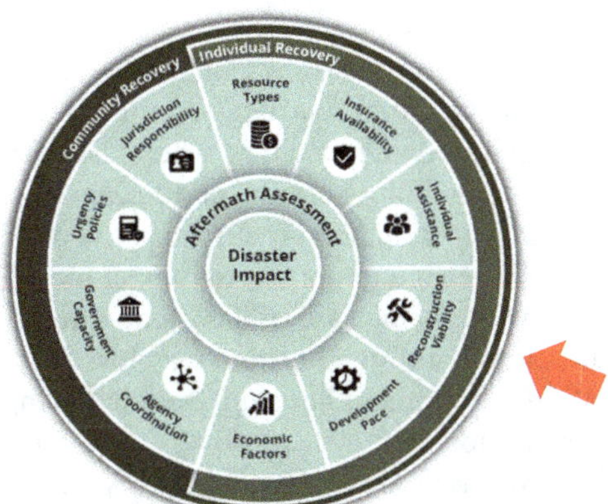

Analyzing community viability centers around the housing stock and market:

- How many housing units are damaged or destroyed?

  - What percentage of the total housing supply is lost?

---

[19] Observed through a variety of arrangements and agreements in the Town of Paradise after the Camp Fire, 2020–2021.

– What is the average age of the home destroyed?
– What is the average price of the home destroyed?
– What percentage is considered affordable?
– What percentage is considered workforce housing?

> What percentage of the lost workforce housing supported trades now in higher demand?

- Is the disaster impacted area known to have a higher degree of title clearance issues, lots lacking in legal determination, lots not surveyed and/or recorded?
- What was the housing supply in the months prior to the disaster?
- Is the housing supply being tracked monthly in all impacted and surrounding jurisdictions?
- Does the cost of rebuilding exceed the value of new construction?
- What is the labor supply in high-demand trades?

  – Is the labor supply local, transient, or imported?

- How has the cost of building materials varied since the disaster?
- Have building codes radically changed statewide or regionally since the disaster?
- What other pressures—national or international—are impacting the supply chain and cost of reconstruction?

Understanding the pre-fire and post-fire housing stock is essential, and several data sources exist to support this. In California, Regional Housing Needs Plans and local government Housing Elements provide statistics on supply and demand prior to disaster.[20] Damage Inspection Reports can aid in understanding the percentage of the impact on the overall housing supply.

Amy Rohrer, Executive Director for Valley Contractors Exchange (VCE), emphasizes the importance of understanding barriers to housing reconstruction early in recovery. VCE regularly gathers data to analyze where, how, and who is recovering by looking at the location of building permits, certificates of occupancy rates, and the percentage of stick builds to manufactured homes. Reconstruction is dependent upon the availability of labor, so it's important to look at the impact of housing loss on essential trades for recovery. In the case of the Camp Fire, VCE's analysis shows that 10% of the construction workforce housing was lost, leaving workers without housing for a period of time.[21] Without housing, even with employment, remaining in the area may not be an option.

If labor is available but materials are in short supply or prices have spiked, recovery may slow and/or proceed at a volatile rate until pricing stabilizes. The impact of increasing lumber prices and building material supply chain issues on Camp Fire

---

[20] Regional Housing Needs Allocation (RHNA), CA Department of Housing and Community Development. https://www.hcd.ca.gov/rhna.

[21] Fellowship interview with Amy Rohrer, Valley Contractors Exchange, 2024.

recovery during the pandemic was clear—projects slowed or halted.[22] Fundamentally, reconstruction can be largely unaffordable if the cost of rebuilding a home constructed 40 years ago exceeds insurance proceeds available to cover replacement value. Any other pressures on the market, including supply and demand affecting home prices, can complicate or reduce the pace of and scale of reconstruction.

Development Pace

Should the local government and community prepare for long-term recovery?

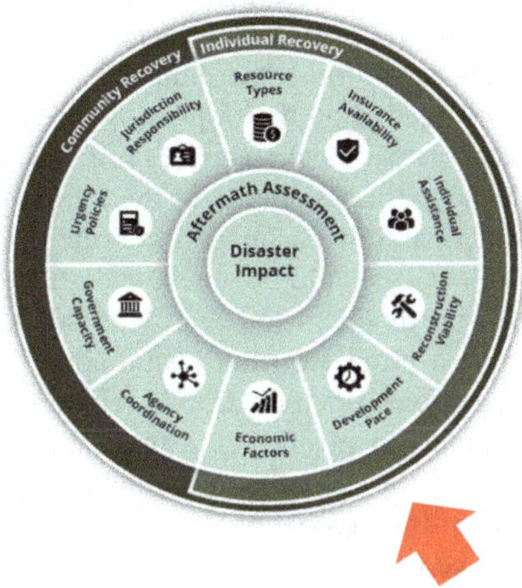

Evaluate the speed at which recovery can occur, and any efficiencies achievable within individual recovery processes, by considering the development landscape:

- Are any impacted areas considered desirable and feasible for master planned developments?
- Are any areas desirable and feasible for contiguous spec building?
- Is the community on the State contractor licensing board's radar for stings?
- What are local attitudes about selling unimproved land to spec builders?
- Are homeowners' attorneys weighing in on the timing of residential land sales?
- Can certain impacted areas support high-density affordable housing with appropriate subsidies, infrastructure, and amenities?
- What is the local permitting capacity and have building codes changed since the disaster?
- Is the State amendable to funding scattered site redevelopment?
- Are mobile home parks impacted and what percentage of park owners live locally?

  - Which parks provide services versus self-service?
  - Is the infrastructure intact for reconstruction, and will the owner rebuild?

---

[22] Observations and anecdotes from fire survivors after the Camp Fire about the impacts of COVID on reconstruction, 2020–2021.

– Are State subsidies available for park reconstruction?
– Can mobile home parks be used as RV parks during the immediate aftermath?

If damaged and destroyed lots are not contiguous, any kind of reconstruction at scale which can be more affordable and efficient for developers, is less feasible. According to VCE, local contractors in Butte County bear the additional high cost of fuel when driving between reconstruction sites in rural remote areas. Contractors must also manage the transportation logistics of moving equipment along private gravel roads maintained by property owners.[23]

Multi-family housing construction requires adequate infrastructure to support density, and available local amenities to support financing.[24] Understanding an area's capacity and regulatory framework for denser housing is key to readiness for grant funding and other subsidies needed by developers. If a destroyed community was infrastructure-constrained prior to disaster, recovery might be the most effective time to increase and improve basic sanitary systems, as the Town of Paradise is proposing to do with the Paradise Regional Sewer Connection Project.

The Paradise Sewer Project, it should be noted, was studied for decades prior to the Camp Fire. The Town went to voters and nearby jurisdictions a number of times to determine amenability to a project of this magnitude.[25] The Camp Fire was a catalyst for the project but certainly not the start of it. Understanding projects needed for a healthy, vibrant community in advance of disaster, through mitigation planning or long-range planning, can lay the groundwork for achieving greater environmental and economic sustainability during recovery.

If a disaster is found to be caused by a corporation and settlements are levied, determinations and pay-outs generally do not come quickly. For this reason, settlement proceeds may not be reliably and immediately available to finance permanent housing recovery. Settlement claims may also be overseen by attorneys who may advise property owners on when and how to sell.[26]

When I was Disaster Recovery Director in the Town of Paradise, a number of multi-family housing developers looking to purchase contiguous lots along main routes near downtown were turned down by property owners at the direction of their attorneys.[27] The instruction was to hold off on selling any property associated with their claim until it was settled. This stymied recovery for those who were not planning to rebuild on their lots, but couldn't yet sell. It also stifled reconstruction for those looking to develop at scale in the most buildable location closest to amenities. Ultimately, though settlement payments are necessary, they add a layer of timing to consider when evaluating resources available for housing recovery.

---

[23] Fellowship interview with Amy Rohrer, Valley Contractors Exchange, 2024.

[24] California Tax Credit Allocation Committee, California State Treasurer web site. https://www.treasurer.ca.gov/ctcac/tax.asp.

[25] History of the Paradise Sewer Project in documents from 1980 to the present. https://paradisesewer.com/resources.

[26] Observations and anecdotes from fire survivors and builders after the 2018 Camp Fire.

[27] Information provided to Town of Paradise staff by prospective builders after the 2018 Camp Fire.

Looking at VCE's data, the Town of Paradise has seen a surge in manufactured housing installation over stick-built reconstruction, largely due to affordability.[28] Community members used to a certain aesthetic may have strong opinions about the type of housing stock returning to their neighborhoods. The availability of manufactured housing can be impacted by similar market forces as traditional construction, as supply and demand determine the timing of installation and the avaibility of installers. It's not uncommon to see a manufactured house sitting on a trailer in a vacant lot for months, while a property owner waits for the installation crew to arrive or the ability to afford installation.

In the months following the Camp Fire, I commuted along the same rural highways as the dump trucks removing debris from Paradise and the Upper Ridge. I'd see trucks lined up in batches of a dozen or more waiting to cross the highway, removing the remains of the place I loved so much: homes, personal belongings, everything. On my way south in the morning, I'd pass a number of trucks hauling manufactured homes north, and know right where they were headed. Modulars, trailers, and pre-framed lumber were also a regular sight. If you know what to look for, recovery is visible within hundreds of miles of a disaster.

Government Capacity

Does the local government have experience and capacity to lead recovery?

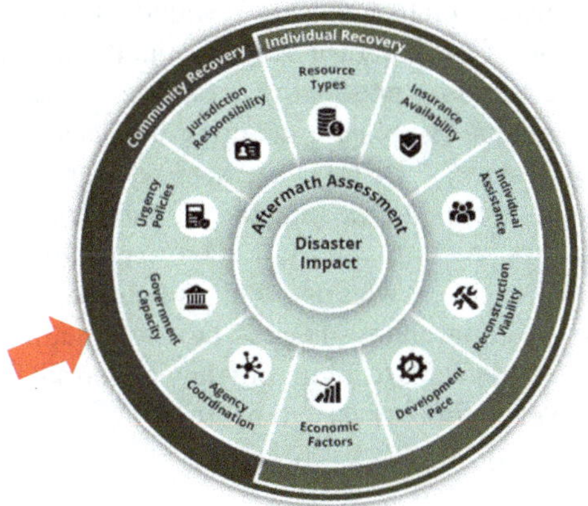

A local government's ability to sustain long-term recovery requires funding, preparedness, and capacity:

- Has the jurisdiction most impacted dealt with disaster before?
- Has the jurisdiction managed FEMA PA funding/projects prior to the disaster?
- Has the jurisdiction managed CDBG prior?
- Has the jurisdiction been given administration authority over recovery funding?

    - Are the funds allocated or competitive?

---

[28] Fellowship interview with Amy Rohrer, Valley Contractors Exchange, 2024.

- Did the jurisdiction have an approved Local Hazard Mitigation Plan at the time of the disaster creating eligibility for cost-share reductions and grant funds?
- Are local/charitable flexible funds available to increase staffing otherwise non-reimbursable through grants?
- What office/department will centralize oversight of recovery planning, projects, funding, and reporting?
- Is there a pre-fire recovery plan, or desire to create one post-disaster?
- What percentage of government workers lost homes in the disaster?
- What are general attitudes toward local, State, and federal government among impacted communities?
- What sources of funds within local governments can be used to front the costs of recovery, and do those sources duplicate federal or State assistance?

Prior experience with recovery funding is helpful but not essential for local governments to administer disaster grants. Local governments learn as they go if they are new to disaster recovery. In many cases, I imagine, jurisdictions deal with small-scale disasters fairly regularly and have at least some framework for understanding what it takes. These can be small storms or coastal surges, or a fire impacting a block or two of a community. Large-scale disasters, however, that are federally declared as they exceed local capacity and resources, do trigger the availability of funding not available for small incidents.

It is important to consider that many first responders and disaster workers are also survivors. In the case of the Paradise Town Council, all five Council members lost their homes in the Camp Fire and had to jump into immediate action to make the emergency proclamations, establish urgency ordinances, and set up alternative facilities for Town governance.[29] Disaster workers who are survivors may have a higher degree of burnout as they are dealing with their personal trauma as well as responding to the needs of the community. Recognizing this from the outset and accommodating personal needs as well as professional needs over the long term is critically important. As the initial chaos subsides, having a trauma-informed approach with the workforce and volunteers, recognizing they are survivors, too, can create a safer, more sustainable space for recovery to begin.

Urgency Policies

Are local policies sufficient to meet changing needs?

---

[29] Shared by Town of Paradise staff following the 2018 Camp Fire.

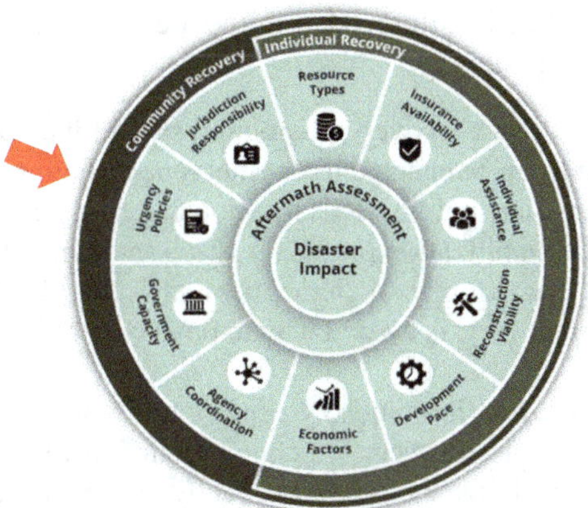

Over time, I've come to consider urgency ordinances necessary safety nets for stabilization and recovery. Urgency ordinances are local policies that create structure and timelines for a series of sequential and co-occurring events needed for recovery to begin, like debris and tree removal, allowable locations for log decks, and temporary housing allowances. Understanding levers built into local urgency policies is essential to identifying and evaluating data to measure their effectiveness and necessity over time.

Understanding the policy framework—the timing of allowances and limitations—and the roles assigned to compliance and enforcement, might include asking:

- Does the jurisdiction have a template urgency ordinance already in use or previously used?
- Specifically, how do existing municipal codes and urgency ordinances differ?
- How do urgency ordinances accommodate survivors' needs inside and outside the burn scar?
- Do State or federal emergency orders have to be layered in, i.e., drought, pandemic?
- What are existing housing policies in and around the disaster-impacted area?
- What are the triggers for housing policy changes?
- What are the pre-disaster policies and permits available for dry camping, RVs, and other temporary housing allowances, and will those meet the needs of recovery and reconstruction?
- What Elements of the General Plan, like the Housing Element, will be affected by the disaster?
- Does the jurisdiction have a Capital Improvement Plan and Design Standards?
- What is the role of the Planning Commission in recovery policy?
- What authorities does the zone administrator have?
- Does the jurisdiction have an adopted Climate Action Plan?
- Does the jurisdiction have the resources to hire and embed a rebuild advocate in planning?

- What was the role of code enforcement in the area pre-fire, and what does their case load look like post-fire?

Disaster planning—and/or recovery planning—should include the identification of template policies that can be enacted in an emergency. Having urgency ordinances drafted to address immediate public needs like housing, and debris and tree removal, can speed up the initial stabilization process post-disaster and clarify the rights and abilities of the public. Knowing what elements of each Board-directed and/or regulatory plan can aid in recovery, is a good way to develop a hub and spoke model for plans to operate together.

Plans are often created during blue sky conditions. Working with consultants to ensure disaster response and recovery are contemplated as plans are developed and updated, is one way to socialize the process of recovery, if not prepare for it directly. Many plans, like Housing Elements within General Plans, are mandated by the State in California.[30] Those Housing Elements must meet certain criteria and standards, and outline required growth in various categories of housing stock.

I arrived at the Town of Paradise on the eve of the Housing Element update which was a fraught but necessary time to relook at how to recover the loss of nearly all the town's housing stock. Reimagining what's possible given infrastructure improvements planned during recovery can be contained within Housing Element updates. Planning processes that recur on a regular basis can be the most efficient way to build toward recovery preparation prior to disaster.

At any given time, the Local Hazard Mitigation Plan, the General Plan, the Housing Element, the Climate Action Plan, the Transportation Master Plan, the Community Wildfire Protection Plan, the Capital Improvement Plan, the Emergency Operations Plan, and so on, may be going through a update. Recognizing these cycles of fine-tuning community planning as opportunities for disaster preparedness and recovery could be the key to including these essential functions in day-to-day work. Granted, many of these plans are led by different departments so interdepartmental coordination is the key to situational awareness and cross-linking outcomes and actions within the plans.[31]

Resource Types

How flexible are the funds for recovery and when are they available?

---

[30] Housing Elements, CA Department of Housing and Community Development web site, https://www.hcd.ca.gov/housing-element.

[31] In Butte County we coordinated the update of the Evacuation Master Plan, Local Hazard Mitigation Plan, Community Wildfire Protection Plan, and Roadside Fuels Reduction Plan, 2022–2025.

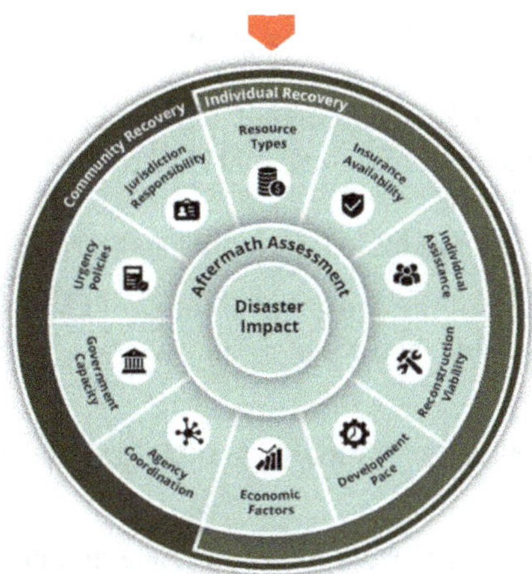

If two federally declared disasters occur at the same time, they may be eligible for the same congressional appropriation of federal assistance. The Camp Fire and Woolsey Fire started on the same day, November 8, 2018. The Woolsey Fire burned 96,949 acres in LA and Ventura Counties, destroyed 1,643 structures, and caused three fatalities.[32] Jurisdictions impacted by the Camp and Woolsey Fires are eligible for the 2018 Community Development Block Grant Disaster Recovery funding.[33] The same funding meant for the same purpose for different disasters in different jurisdictions may yield wildly different outcomes due to capacity, infrastructure, priorities, and other variables.

Non-government organizations (NGOs) may also be eligible for federal assistance. In my experience, it takes years for local governments and NGOs to settle into their respective recovery roles. Early on, it may be helpful to scan local organizations for funding capacity and expertise to take on recovery roles. Keep in mind NGOs may be called into service during response.

Disaster recovery professionals I spoke with in Oregon following their major firestorms after the Camp Fire, reported a crowded and overlapping pool of NGOs trying to establish post-fire roles and responsibilities.[34] This confusion led to hardship for well-meaning individuals attempting to find their lanes and gain traction with support and resources. In my opinion, avoiding this is likely impossible, but anticipating it with pre-planning might loosely assign roles that give everyone a place to start.

Framing out the availability of recovery resources could start with these questions:

---

[32] CAL FIRE Incident Report, Woolsey Fire, last updated 7/7/2025. https://www.fire.ca.gov/incidents/2018/11/8/woolsey-fire/

[33] State of California 2018 CDBG-DR Action Plan, CA Department of Housing and Community Development, August 2020. https://www.hcd.ca.gov/community-development/disaster-recovery-programs/cdbg-dr/cdbg-dr-2018/docs/HCD-CDBG-DR-2018_AP-Final-ADA-English.pdf.

[34] Anecdotes shared by disaster recovery professionals working in Oregon, 2020–2021.

- Does the fire meet the threshold for a state of emergency and/or presidential declaration?
- Has FEMA IA and/or PA been authorized?
- Does the jurisdiction have a compliant Local Hazard Mitigation Plan for cost-share reduction/waiver eligibility?
- Are financial settlements available for victims and/or responding agencies?

  - If so, what is the size of the Trust and number of claims?
  - Are settlements taxable and what is the average percentage deducted for attorney fees?

- What local funding sources are available to rebuild, i.e., community foundations, charitable organizations?

  - Which local organizations have the experience and capacity to scale quickly for response and recovery?
  - Are local organizations prepared to apply for and manage State and federal recovery grants?

- What is the availability and timing of any federal/State assistance for response and recovery costs?

  - What agencies are contracting debris and/or tree removal?
  - Are debris and tree removal programs consolidated?
  - What properties—commercial, residential, size, etc.—are eligible for a debris/ tree removal program?
  - What agency collects insurance proceeds for debris removal reimbursement?
  - Is there a match or local cost-share requirement for any response and recovery actions?
  - Are government settlements considered duplication of benefits?

Funding available for recovery comes in different timelines tied to specific uses. Understanding the differences between first-in money (insurance/FEMA) and last-in money (CDBG-DR) will allow jurisdictions—and individuals—to avoid costly mistakes in handling insurance and/or FEMA claims. Charitable funds can and should be considered the most flexible and may fill gaps when local matches are required for State and federal grants, and/or meet needs otherwise ineligible for disaster recovery funding.

After disaster, impacted individuals may experience a heightened state of trauma and may not be able to retain the complicated matrix of funding information coming their way. I say this because I observed my parents going through the process and felt their acute information overload. Misunderstandings about how to use funds in early recovery can be detrimental if not downright catastrophic to individuals and households as recovery progresses. Clear and accurate local public messaging can help fire survivors avoid financial mistakes, if regulatory agencies produce Frequently Asked Questions documents about their programs.

Before giving charitably on any scale, it is important to understand what documentation survivors must provide to qualify for government assistance which, if available, often represents the bulk of funding for recovery. Charitable giving is well intended, but if the timing and method of disbursing relief decreases an individual's ability to secure critical funding for reconstruction or long-term housing, those charitable funds may—counterintuitively—work against recovery.

Managing funding for recovery involves the complex art of understanding the sources and uses of each funding type, and weaving them together to meet community needs. Avoiding duplication of benefits and having proper back-up to justify expenses are probably the most critical steps to maximizing resources for recovery. Several years after the Camp Fire, we are still sorting through what constitutes duplication of benefits for some programs, and producing back-up documentation for reimbursement. Grant funding available for mitigation must not duplicate the good works done for free on private properties by non-profits. Grants must be contemplated for entirely separate processes than funding already secured for the same purpose.

Individual Assistance

What percentage of survivors will need outside assistance and resources?

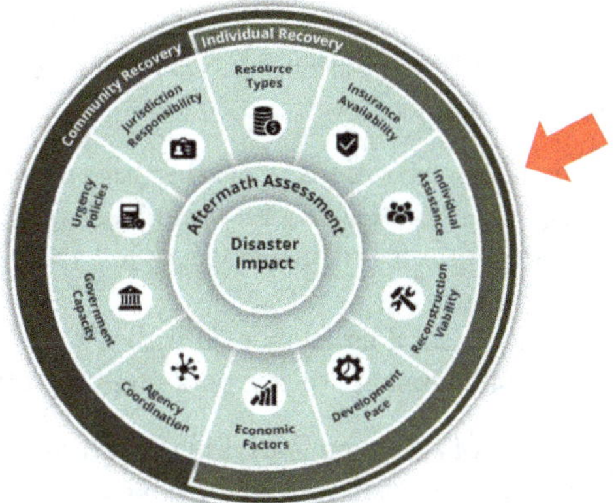

Narrowing in on resources available to disaster-impacted individuals is key to understanding the pace and scale of individual recovery. Many fire survivors in the Butte County area lack adequate resources to reconstruct their housing without assistance.[35]

It's essential to piece apart the resources survivors have access to, and the various expenses related to recovery. It's not uncommon for survivors to use all insurance funding and federal assistance, if applicable, on preparing their lot for reconstruction, only to find they lack the resources for materials and labor to rebuild. Or, for renters to use savings to purchase a vacant lot with every intention to build, only to find

---

[35] Self-reported data provided by fire survivors during intake at the Local Assistance Center for the Park and Thompson Fires in 2024 indicate 45% of homeowners had no insurance, 55% had some—often inadequate—insurance.

they're unable to secure a loan or other financing for construction. Funding recovery on the individual level can be similar to putting a puzzle together—every piece must fit perfectly for the process to work.

If a disaster does not meet the federal declaration threshold for FEMA assistance but the damage is significant, State grants or loans may still be available.[36] Smaller fires may not trigger any government assistance at all, however, and survivors must rely on any insurance proceeds, remaining assets, the ability to finance a permanent housing solution, the aid of friends and family, and/or temporary housing. Funding availability has the broadest implications for how quickly and completely individuals can recover.

Supporting individual recovery starts with understanding:

- Does the disaster qualify for FEMA Individual Assistance (IA)?

    - If yes:

        How many IA claims were submitted by survivors versus funded?
        Where is the Local Assistance Center or FEMA IA assistance center in proximity to the impacted area? Is additional transportation needed to facilitate access?
        What is the placement of emergency government housing in relation to the impacted area? Are FEMA trailers placed on impacted private properties or well outside the burn scar?

- Are securely funded local organizations providing long-term case management?
- Is there an unmet needs roundtable attended/organized locally?
- What is the timing and eligibility of owner-occupied reconstruction grant/loan programs?
- Does the State plan for any rental assistance/subsidies?
- What is the space count and capacity of RV parks within 100 miles?

Digging into the process of securing individual funding is essential for local governments to identify what barriers they might be able to remove, such as lowering permit costs. Understanding the limits of FEMA Individual Assistance, and permitted uses of the funds, can help identify gaps that might be filled by charitable funds or government loans and grants. The timing of those government loans and grants may delay the ability of survivors to reconstruct permanent housing or purchase a home outside of the burn scar. Because there is no centralized system of recovery where funding decisions and disbursements are coordinated among agencies, understanding the ins and outs of recovery funding for individuals may help address gaps and reduce barriers. Individual case management may help survivors navigate and piece together enough funding for recovery.

---

[36] California Disaster Assistance Act (CDAA) funding can be authorized by the Governor in the State of California. https://www.caloes.ca.gov/office-of-the-director/operations/recovery-directorate/recovery-operations/public-assistance/california-disaster-assistance-act/.

Agency Coordination

Are the right agencies working together cooperatively?

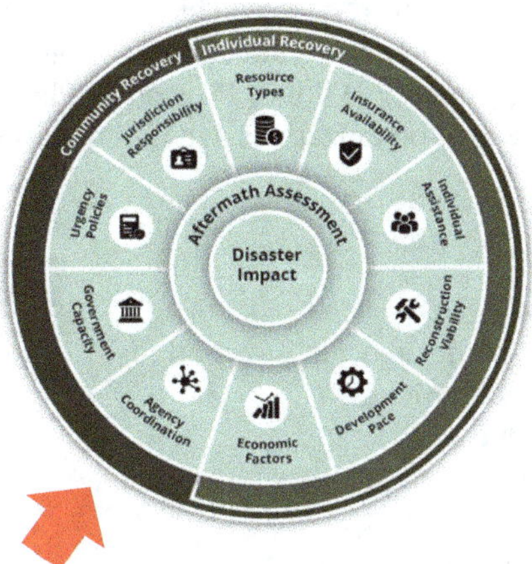

Agencies, governments, non-profits, elected officials, membership associations, and the many other organizations involved in recovery all have different roles and responsibilities. The ability of these groups to pursue complementary goals for the overall benefit of the recovering community depends upon clear, transparent communication and coordination, and an understanding of public and private sector differences.

An issue I observed working for the Town of Paradise in early Camp Fire recovery was a blurring of boundaries between organizations. If organizations are collaborating to achieve the same mission, messaging should be coordinated so there's no wrong door. If organizations are attempting to achieve similar outcomes without communication or cooperation, survivors may be unclear where to find support. As organizations find themselves in helping roles and learn new duties and responsibilities, it is important to consider messaging and expectation-setting for larger audiences, including governmental agencies, to ensure duplication and/or conflict does not occur.

Disaster creates opportunities for leadership, and it's natural for individuals and organizations to want to be part of response and recovery. In fact, I've heard some community leaders explain their personal involvement in recovery is their "trauma response" to disaster.[37] It is critical, however, to consider that ownership and leadership over recovery should lie with the impacted community, not with those standing in for survivors just because they are available. Even meeting locations matter— impacted communities may balk at leaders from other communities convening on their behalf to solve their challenges. The closer recovery can be coordinated to the epicenter of disaster, even amid the destruction and challenges, the more likely

---

[37] Observations and anecdotes from disaster workers within Butte County following numerous fires, 2022–2025.

support and resources will be accepted by survivors and utilized toward recovery goals and objectives.

Before disaster, get to know organizations and individuals who might find themselves responding to or aiding in recovery. This can be accomplished by getting involved in various planning processes like General Plan Updates, Local Hazard Mitigation Plans, and Community Wildfire Protection Plans. Familiarity and trust may reduce the initial chaos of organizations and individuals stepping on each other to assist.

Build trusting relationships with:

- Federal, State, and local tribes
- Staff and elected officials in adjacent jurisdictions
- Federal and State elected officials
- Federal and State lobbyists
- Board of education and school districts
- Utility companies
- Property Owners Associations
- Chambers of Commerce
- Downtown Business Associations
- Realtor Associations
- United Policyholders/Department of Insurance
- Special Districts—water, wastewater, parks/recreation, tourism
- Statewide county and city associations
- Land stewards, i.e., Associated Governments, Resource/Conservation Districts, Fire Safe Councils.

Recovery should respect and honor boundaries in the way disaster cannot, but it should be cooperative and collaborative if possible. When local governments, districts, agencies, and non-profits collaborate on emergency planning, mitigation planning, evacuation planning, and other critical efforts that can be documented and table-topped, they become familiar with decision-makers and their functions. Even though jurisdictions are largely forced to manage recovery on their own given the ways in which funding is allocated, information sharing through planning (in advance of disaster, if possible) should be a central value for guiding regional recovery. Specific to the Camp Fire, nothing pains me more in long-term recovery than discovering the Town is suffering alone through something the County can aid in, and vice versa.

A big chunk of time in recovery is spent on advocacy which relies heavily on the support of elected officials at State and federal levels. There are constraints on both individual and community recovery. Understanding and being able to articulate various constraints to decision-makers is critical. Personally, I believe if we can't change our recovery for the better, we can make barriers known to change outcomes for other communities impacted by disaster in the future.

Logistically, survivors can be difficult to reach after disaster for a variety of reasons. Organizations that have built trust and established relationships in impacted communities may have more success messaging out programs and services than government agencies.

We're now several years into recovering from the Camp Fire and we still rely on trusted community organizations to aid in planning, programming, and messaging.[38] A barrier the State regularly faces during recovery is a lack of trust among the public, therefore, willing local community groups may be the best facilitators for information sharing. An important caveat here is for each organization and agency to carefully avoid making promises and setting expectations on behalf of any other entity.

I learned early in my time as Disaster Recovery Director that the messenger will be vilified if information is false or guarantees are not met. I was the trusted voice of recovery and regularly presented updates to the Town Council on behalf of Cal OES and CalRecycle related to the Hazard Tree Removal Program. I'd receive State updates on daily or weekly calls and present them to the Council and public, only to find out later major shifts were occurring in the programs unbeknownst to the Town.

As the voice, I would have to navigate questions and, in some cases, understandable fury. I remember being shouted at while delivering a presentation, then taking a break in the hallway only to be threatened by a member of the public.[39] This frustration and anger at the delays and inconsistencies were fair, but as the person appointed to absorb it all, I found it challenging. Thankfully our staff and elected officials were largely supportive and the weight was shifted and shared amongst staff during particularly heated topics and moments. Eventually, I learned it was better for the agencies to speak for themselves and stepped out of the way.

Emergency response systems like the National Incident Management System has very specific protocol for communications such as the requirement to use plain language.[40] To the extent disaster recovery professionals can avail themselves of those emergency management practices, the more controlled public messaging can be. During the chaotic early phases of response and recovery, it is challenging to know how all information will land with elected officials, the media, and the public. Relying on public information professionals who know how to address challenging topics is key to maintaining clarity through the long process of recovery.

---

[38] Examples include the Butte County Resource Conservation District and Butte County Fire Safe Council, both part of the Butte County Collaborative Group formed after the 2018 Camp Fire. https://butte-county-collaborative-group-bcrcd.hub.arcgis.com/.

[39] Town of Paradise staff initiated safety protocols for situations like these.

[40] National Incident Management System, Third Edition, October 2017, FEMA, page 24. https://www.fema.gov/sites/default/files/2020-07/fema_nims_doctrine-2017.pdf.

Chapter 8
Family

My oldest daughter's birthday is on December 31st and my grandma sent her a birthday card every year. Just after Christmas in 2014, after we'd spent the holidays doing puzzles with grandma in Chico, we got a call from her home in Petaluma that her knee was hurting severely. The events of the next few hours are blurry but she went from an ambulance for her knee into heart surgery. I was on my way to Petaluma with my mom in the front seat and Adela in the back seat, ready to help grandma recover from surgery, when my aunt called and said she had died on the operating table.

I found a rare exit on I-5 and while my mom cried over the phone with her sister I pulled onto the gravel. I knelt behind my car and called my husband who talked me through my hyperventilation and got me back on the road. I delivered us to Petaluma a few hours later. We spent several days in grandma's house with my aunts and uncles who'd flown in from across the country, and sat together in mostly stunned silence and reflection, save for the kinds of arguments siblings have when an estate is at stake and no one is quite in charge yet.

We drove home to Chico in early January to let the dust settle and to get back to work and school. In our mailbox we found a birthday card for Adela from grandma, a thick pink envelope with a sticker on the outside. We wept as we read grandma's trademark cursive writing wishing Adela a happy year ahead. She never missed a birthday, even after she was gone.

I heard similar stories from fire survivors after the Camp Fire. A friend arrived at his lot to find his house destroyed but his mailbox intact, a card inside from grandma to his kids. The fire had blown right over it. It wasn't lost on me that the card we received at our house was from a grandma who was gone, and the card my friend received was from a grandma who was very much alive to a house that was gone.

When I worked for the Town, the mailman in Paradise was several decades into his career and still as dedicated as ever to his route. With most of the houses gone now, he remarked how easy it was to find porta-potties when he needed them on the job, with several lots under reconstruction along his way. I remember studying his face and the way he carried himself for any hint of trauma, and I couldn't find a single one. He was proud, working, and excused himself to get back to his route.

We still rely on mail for recovery and emergency planning because many households in remote areas are underserved or unserved by broadband.[1] We think about technological redundancy for emergency communications, and about redundant forms of communication for day-to-day announcements and activities.

In many small, rural communities in Butte County, the tried-and-true method of putting up sandwich board signs at local grocery stores and community centers is often the best and only way to get information to the public.[2]

8.1 Recovery Shuffle

Wildfire traumatizes a community all at once, leaving everybody reeling with shock and dealing with one another and themselves in a heightened emotional state. There is no soft roll of trauma through a destroyed community, rippling gently from person to person, there is only one loud, sharp, invasive, explosive experience.

I have noticed in myself and others that even friends and neighbors with shared histories and companionable relationships can grow apart rapidly depending on how differently the fire impacted them. For me, I have carried a hard, heavy heart since joining the disaster recovery profession and have found it nearly impossible to relate to anyone who doesn't do this for a living.

In 2023, I remember standing with a friend from junior high at a bridal shower in the Bay Area, listening to her talk about "pod'ing" up with neighbors through COVID, and feeling like I was a million miles away. The only thing I could think to share was how I spent the pandemic working in disaster recovery while experiencing another fire, but that didn't seem like proper party etiquette so I kept quiet.

A co-worker at the Town expressed her extreme survivor's guilt upon finding her house still standing after the fire. Unable to celebrate and imagine her life without the people she'd lived around for years, she relocated to another standing home in the town to start over. On the flip side, friends who'd lived apart pre-fire RV'd together post-fire, and wound up building their homes on adjacent lots to stay close.[3]

To me, wildfire is a hand grenade thrown recklessly at neighborhoods, friendships, family structures, routines, traditions, everything. When people and places and things begin to resettle after the smoke clears, they often reshuffle into new configurations. I now receive Christmas cards from Minnesota, Iowa, and Oregon from friends who used to live right down the street. In Paradise, it's not uncommon for families who

---

[1] Butte County Broadband Planning and Feasibility Study, produced by Tilson Technologies, adopted by the Butte County Board of Supervisors o February 27, 2024. https://www.buttecounty. net/DocumentCenter/View/14226/Butte-County-Broadband-Planning-and-Feasibility-Study? bidId=

[2] Information provided by Jim Houtman, Butte County Fire Safe Council, 2024.

[3] Anecdote from Melissa Schuster, Judy Clemens, Pam Hartley, and Kelley Connors following the 2018 Camp Fire.

lose beautiful homes on steep hillsides with great views to buy lots across the street to rebuild on flat ground. Post-fire reshuffling can be monumental or incremental, and it's rare for things go back to the way they were.

After fire and reconstruction, homeowners desiring more control over nearby vegetation and defensible space may purchase vacant neighboring lots to protect their re-investment and their sanity. Ultimately, Paradise may end up with a smaller population than it had pre-fire, in part for this reason.[4]

A family who resettled in Chico drove their kids to and from the high school in Paradise to keep them with their friends who were also being driven in from elsewhere. Paradise High School survived and was the only familiar space their kids shared with friends anymore. Their son starred on the football team after the fire and gave his family a reason to cheer for Paradise.

My parents had several housing options following the fire but chose to remain in Butte Creek Canyon just down from Paradise even though the land's demands and their mobility are changing in inverse proportions. With each season, their property needs more from them, though the majority of their trees have been removed or fallen, often barely missing their house. The vegetation on their property grows on a rocky hillside behind their house and requires regular hand-cutting, hauling, and pile burning, otherwise they're surrounded by fuels. Their land lacks the shade it once had with the tree canopy, and after rainy winters the spring grasses are knee-high.

As I process my own experiences and get some distance from the Camp Fire, I am only just beginning to see the magnitude of pressure it put on our community to get along, get by, and get ahead in the aftermath. Relationships formed during response and recovery – trauma bonds – can turn intensely fraught as time goes by and circumstances change. Dependence upon one another during uncertainty can flip on a dime to mistrust and suspicion when there's no shared villain, as exhaustion and rage set in.

At the five-year mark after the Camp Fire, organizations and governments scaled back or shifted recovery staffing associated with the Camp Fire, perhaps for the last time, setting off a final wave of retirements and employment shuffling as another round of new normal sets in.[5] I am not the only local professional who has experienced one or more job changes related to fire and recovery. Many colleagues have not only changed jobs but changed industries, jumping head-first into a new learning curve as I did.

After a major wildfire, life churns for a good long time which is something I wish I'd seen coming. I might have given myself and my parents more grace if, as I watched the flames bear down, I prepared for frenetic decision-making, terrifying illnesses, sweeping life changes and reversals, friendships born and broken, and career upheaval. If I'd known what was ahead as the firefighters gained control and the emergency subsided, I might have entered the aftermath with greater preparation

---

[4] Observations and anecdotes shared by fire survivors after the 2018 Camp Fire.

[5] Observation of a number of non-profits and private corporations five years after the 2018 Camp Fire.

for the chaos that had only just begun, instead of expecting the opposite. Shock makes the wall of recovery harder to climb.

8.2 Paradise Magic

I've noticed two things occurring as a result of writing this book: (1) jotting down my memories allows them to loosen their grip on my mind; and (2) my grief is fading. For me, writing is the ultimate act of letting go, shifting things lost in my mind to found on a page.

In 2024, I attended a tourism strategic planning session at Melissa Schuster's home in Paradise. Melissa was the first friend I made when I moved to Paradise in 2008. She was on the Paradise Ridge Chamber of Commerce Board of Directors when I was hired as Executive Director, and took me under her wing.

Melissa's property survived the 2008 Humboldt fire but just barely. She lives in lower Paradise and arrived home to find fire retardant dumped all over her property and the fire line within view. She let firefighters shower and sleep in their guest rooms after slogging through long days and nights. I'll never forget Melissa calling to tell me her property had survived. I was standing on my mom's porch in Chico having evacuated Paradise a few days earlier with Adela. I was still brand new to Paradise but I cried with pure joy and relief for her.

Before the Camp Fire, Melissa's property was a stunning, sprawling estate. It had a Manor House she operated as a three-room bed and breakfast, the Rock House where she and her husband lived, the Pavilion where they hosted outdoor barbeques and events, the garage that held their classic car collection, an historic three-story watch tower held together by vines, a gazebo overlooking the pool, a cave with a dance floor and bar, a chapel where she decorated Christmas trees every year, a koi pond, a gate house at the front of the property, and a two-story barn next to the llama field. After the fire, only the chapel, the cave, and the pond remained, an eclectic mix of structures for Melissa to build her recovery vision upon.

In the years following the Camp Fire, Melissa constructed three themed glamping tents overlooking the pond, a vaulted double-decker gazebo, and a two-story fire resilient home with an elevator for aging in place. On the slope leading up to her house on the day we visited, at least 100 mama and baby goats munched the wild green grass. At one point a tiny goat with a curly brown coat got stuck on the wrong side of the fence and stood crying for his mama. Melissa hopped in her 4-wheel-drive golf cart with a couple of volunteers and reunited the pair.

Down the hill from Melissa's house is a brand-new steel-framed barn where her son is distilling spirits. While giving a tour of the distillery and tasting room, her son explained the last step in obtaining their operating permit is installing parking spaces in the gravel lot out front. The heavy lifting is done and they are nearly ready for tasters under the twinkling chandelier.

Melissa explained they purchased adjacent lots for a total of 18 acres. Past the porch of the tasting room is a driveway and the remaining brick façade of her former neighbor's house. The property now belongs to Melissa who is thinking of building a restaurant and event space in the scenic ruins.

Up the hill off the gazebo is the pond where evacuated koi fish joined Melissa's fish during and after the Camp Fire. As we stood over the pond watching the fish eat, we saw two turtles swimming mightily toward the sinking food, the koi spinning them out of the way. Catfish live in the depths of the pond and appear when there's food on the surface. Years ago, when Adela was no older than a baby goat herself, she would sit on Melissa's island and feed the catfish tiny brown pellets from a can.

Long before the Camp Fire, I knew Melissa could make magic. She had vision and the ability to see her vision through. Now, walking her property listening to her tell the story of their recovery, I felt like I was witnessing a miracle. Melissa's son explained that just weeks after the fire his mom began lobbying him to move back to Paradise from out of state to open the distillery, his dream business. He was reluctant fearing there was nothing left in Paradise, but she eventually convinced him. He and his wife and kids moved back to Paradise and together they've created something special because Melissa had the inspiration, the iron will, and the means to make magic happen.

I told my boss after the visit that it was the first time since the Camp Fire I'd been able to celebrate something new in Paradise without missing what it replaced. Many of Melissa's trees are healthy oaks and redwoods, but several others are weak and pained. The fire is visible and present in its quiet, constant way, but Melissa's own healthy resilience, her positive attitude, her sheer delight in her home and her property, the adventure she is clearly on in re-creating it, add up to more than the fire took away.

Melissa shared on the tour that she and her husband had only to look at their property once after the fire left it stark and ashy to see it was enough. Nothing but blackened earth for miles around – and it was enough.[6]

8.3   Theater on the Ridge

Adela has the same soft spot in her heart for Paradise as I do. When I worked at the Town, my dear friend Judy Clemens, another esteemed Paradise lady, reached out and invited Adela to intern at the Theater on the Ridge. Judy anointed her "Assistant Director" knowing how well it would suit Adela's personality and motivate her to lead.

Judy runs Theater on the Ridge which was founded in 1975 and is the oldest non-profit community theater north of Sacramento. The theater has 101 seats and runs a few shows a year.[7] Miraculously, the theater survived the Camp Fire months after the building was paid off.[8]

---

[6] Information, stories, and tour provided by Melissa Schuster, 2024.

[7] Theater on the Ridge web site, https://www.totr.org/blank.

[8] Anecdote provided by Judy Clemens, one of the founding members of Theater on the Ridge, 2021.

After school and in the evenings, Adela would drive up to Paradise to help at the theater. She'd just gotten her license so this drive up and down Skyway was a big deal. The play, the Wild Women of Winedale, starred a number of women just like Judy who had been acting for years, and they took on Adela as their own.

As the production got closer to opening night, Adela would dress all in black so she could scoot around the stage moving props and setting scenes without being seen. She learned how to work the curtains and on opening weekend we went up to see the play.

Judy gives the best hugs in the world. She has the warmest, strongest, most complete embrace that you feel in your heart. After the play we visited with Judy before heading down the hill. A few weeks later, the ladies treated Adela to a bag branded with the show's logo and her name embroidered on the side, a wrap party gift. The stars of the show found me on Facebook and raved about Adela.

I moved to Paradise when Adela was three years old, and by the time she was four I was desperately seeking support for parenting such a strong-willed child. One of my mom's friends told me I had to grow faster and farther than Adela so I could always stay one step ahead. That advice never left me.

At five years old, Adela would lay awake at night in her pink canopy bed in Paradise and talk about "house practice." She had big plans to move into the vacant house next door with canned food and plenty of clothes so she could practice living on her own. She was adamant: without house practice she'd never be ready to move out for real which she longed to do. I'd sit and listen to her, exhausted from the day, wondering if other kindergarteners were making similar plans to move out.

In those days, Judy always answered her phone for me no matter what time of day or night. She was my lifeline well after dark when Adela would argue with me for hours about what she'd prefer to be doing besides going to sleep. I'd pace on the lawn in the dark under that beautiful, mature shade tree in the front yard, and Judy would coach me through the moment.

We didn't necessarily lose touch after the Camp Fire but we didn't connect regularly outside of social media, so it was a blessing to have Adela work with Judy before graduating from high school and moving on. I had no idea when I lived in Paradise or even when Adela worked at the theater how quickly her childhood was disappearing. Judy got to have her moment with Adela, Adela got to have her moment in Paradise, and I got to have my moment with Judy. It was fleeting but healing and reminded me that love persists, just like the strong wills that make strong women who rebuild their lives.

8.4 Three Fire Mountains

Toward the end of the fellowship, Heidi looked over my shoulder and asked why I was writing about the fire when it happened so long ago. To her, it was literally half a lifetime ago, to me it was like yesterday. I told her I needed to get my memories down so I could finally let go and move on.

I asked Heidi what she remembers about the Camp Fire. She said she remembers her ears popping while we were evacuating. She remembers looking up at school that day and seeing the black sky. She fell down when she was looking up and a friend asked if she was ok. After telling me this story she paused and said, "I remember seeing three fire mountains in the background," and then she left the room.

Because we didn't lose our home and her school and our parks and grocery stores in the fire, Heidi was able to watch it from a distance. She felt the aftermath, though. Her classmates and school environment changed overnight. Friends who lost homes in Paradise moved away, kids from Paradise schools joined her class, Chico families that lost jobs in Paradise left the area, kids left then came back years later.[9] All part of the great wildfire shuffle.

After we evacuated to Sacramento the first night of the Camp Fire, Heidi and her cousin, Violet, went to Lake Tahoe with my parents to stay at my aunt's house away from the chaos. The air was thick and heavy with smoke even in Sacramento so they drove into the Sierras where the wind blew in their favor. The devastation in Butte County was catastrophic and there was nothing my parents could do by staying in the area. Information was trickling in and it was hard to know what was true. Over the course of the first week, they heard conflicting stories from friends and neighbors about their property and other nearby homes. Tahoe represented a much-needed breath of fresh air from the fire which burned and smoked for another two weeks.

My parents did their best to make the trip feel like a vacation. They kept the mood light for their granddaughters who were disoriented from being swept out of their homes, schools, and community without warning. I drove up to Tahoe to join them at one point and remember leaving the heavy smoke for the first time in over a week and being able to roll down my windows and inhale. I pulled over so I could stand in the cool clean air and blue sky, a bit disoriented myself.

I'm not sure we could have handled the kids' experiences any differently. The fire was clearly visible from their schools and once we got them out of the community, we did our best to process with them in age-appropriate ways. I think I locked my grief up tightly until just recently when I started grieving the cows, so I'm fairly certain life went on as best it could.

I've thought about Heidi's question a few times, wondering if it is time to let go of the fire. I can feel a willingness to let go of the grief without feeling like I'm betraying Paradise, but I don't feel ready to let go of the fire itself. My job is to see disaster recovery funding through, so in many practical ways the fire still defines my working life. It is a memory I keep somewhat fresh in part to be ready for the next Emergency Operations Center activation, and in part to honor the reason for our work.

I am terrified of another fire hitting our county. We refer to intact communities in our wooded foothills as those that "haven't burned yet" as it's only a matter of time, but I sincerely hope we're wrong about that. As much as we distract ourselves with

---

[9] Observations and anecdotes from friends, family, and fire survivors for several years following the 2018 Camp Fire.

the busy work of life, I feel another threat lurking around the corner. I think this has strained my peace of mind as much as the heartbreak itself.

Over the past year we have planted seven baby trees in our backyard: a lemon, pomegranate, ash, maple, mulberry, redwood, and oak. We didn't do this as a direct result of the fires as we added them one by one, but I have no doubt adding tiny trees to our lives is filling a hole in my heart left by the removal of all those giant beauties on the ridge.

I wish healing was as simple as feeling the hurt from the fire, addressing it, and moving on, but it's not working out that way. Instead, it's a layering of good things on top of new memories on top of mental and emotional efforts to move past common reactions to trauma, in hopes the cycle doesn't repeat again. I can only measure the depth of my grief by the volume of goodness I pour into it.

Heidi has memories of three fire mountains bearing down on her school and that's all she wants to say about it. My hope for her is that even if she continues to experience these disasters, she keeps her joy, her positivity, and her ability to wonder why anyone would carry fire longer than they have to.

8.5 Fire Stories

My husband was reluctant to talk to me about his experience with the Camp Fire. It wasn't until I asked him about it for this fellowship that I realized we'd never really dug into the details before. When the fire happened, we talked about the logistics of evacuation and our kids' schedules and whereabouts, but never really about his decision to stay home.

He remembers three things from that day: the unusual direction of the wind, the speed of the wind, and the black smoke that came over Chico in a way it never had before. Adela was the first to alert him that something wasn't right. He ran outside to see the same smoke shelf she was seeing from school moving slowly over the sun like a fire eclipse.

I asked him why he isn't traumatized from the fire and he said he is in his own way. For days, weeks, even months he immersed himself in social media and the police scanner, consuming every piece of information as it was happening – right and wrong. At one point, reports said houses just east of us were burning but they weren't.

Looking back on our lives over the past several years, I see the toll of his fight and my flight during the fire.

After the kids and I drove away that night, he and our neighbor made a pact: as soon as one house in the neighborhood went up, they would grab our elderly neighbor and leave. As we drove toward I-5, he watered the lawn, the fence, the roof, the patio, everything. He could see flames over the rooftops in the distance and, as soon as the sun set, the size of the fire was unmistakable.

Our house is on a corner up the street from an elder care facility. He said at one point twenty ambulances blasted past on their way to deliver patients from Paradise.

He counted twenty-four fire engines stationed in a hard line at the edge of a field up the street.

When I told him I was considering the job in Paradise he was concerned. I'd just established myself at the organization in Sacramento, the start of a brand-new career chapter, and was already contemplating change as a result of the fire. Like many people, he was unsure about the viability of the town and didn't have much information to go on. Neither of us knew what disaster recovery would entail, nor did we have any inkling how many years I'd be involved.

It took writing more than 100 pages of this book to ask Heidi and Rodney about their fire experiences. It's interesting to me that I talk about fire non-stop at the office yet rarely at home. I think this compartmentalization is not uncommon amongst households still dealing with the aftermath to this day.

Rodney has been through a number of fires having lived in Chico his whole life. Fire recurred every 10 years or so in this area prior to the last several years when mega fires happen back-to-back-to-back.[10] He says in all his decades here, in all the fires he's seen, the Camp Fire was by far the scariest.

Nearly everyone in Chico has a fire story. Many people who were not directly impacted jumped into volunteering as the fire burned, serving food, sorting donations. Certainly everyone on the ridge has a fire story, maybe more than one. Our married friends evacuated separately with their kids without cell service hoping to find each other at the bottom of the hill. Somewhere. Anywhere. Alive.

As I write this collection of fire stories from my family, I wonder how relevant our Camp Fire experiences are to the rest of the world. I wonder about the value of offering details of our recovery if they never matter to anyone other than us. But as I said to my fellowship mentors, if I can shortcut any lessons and realizations learned the hardest way in disaster recovery, then I've done my job. If I can help one disaster recovery professional normalize the inevitable chaos and confusion, then this was worth it. I'm certain not everyone will agree my family's fire stories are worth telling, but it seems the very essence of the human spirit to offer something heartfelt to those who are hurting.

8.6 Homemade Home

Before I moved to Paradise in 2008, I was a convention director for an international medical association headquartered in the Bay Area. Planning meetings and conventions all over the world meant a lot of travel – planes, trains, and taxis. On one memorable 9-day trip I had meetings in Seoul, Hong Kong, Kuala Lumpur, Singapore, Brisbane, and Melbourne. I started the job one week before September 11, 2001, and arrived to work that morning to find everyone huddled around a tiny plug-in tv on the conference table.

---

[10] Information provided by Jim Broshears, Retired Fire Chief for the Town of Paradise, 2024.

Our association membership was made up of medical doctors, medical students, physicians, clinicians, radiologists, and technicians from nearly every country. Even the President of Harvard was a member at one point. My job as convention director was to plan and manage the logistics of scientific conferences and technical exhibitions that rotated between the east and west coasts of North America, Europe, and Asia.

Between 2001 and 2008, I planned dozens of events: a workshop on a cruise ship, a meeting during a hurricane on San Marcos Island in Florida, and a seminar in Brazil just one week after Carnival. In one ten-month stretch we moved a meeting from Barcelona to Berlin, cutting the six-year planning timeline down by five years and two months. I've had food poisoning in some of the most exotic cities in the world, climbed the Eiffel Tower, and learned not to schedule a layover in Frankfurt any shorter than 2 hours to account for the running in between flights. Because I walked with confidence everywhere I went, I've been asked for directions in almost every language you can imagine. I embraced the awkwardness of travel which included unpacking my entire suitcase on a moving train during rush hour in Austria to find my ticket in a coat pocket, and getting stuck and shouted at in a subway turnstile in Spain. My passport is a celebration of my early career.

I oversaw the construction of mini cities in some of the world's largest exhibit halls with subfloors, plumbing, electrical, overhead lighting, pipe and drape dividers, stanchioned coffee breaks, and two-story booths. These cities were constructed to facilitate gobal exchange between doctors and MRI technology and equipment manufacturers. I was very proud of that work then and now.

Before relocating to Chico, I applied for a number of jobs in the area and was called back for an interview with the Paradise Ridge Chamber of Commerce. The Town Manager was on the hiring panel and the retiring Executive Director showed me around town. There was a written test as part of the evaluation and, after several conversations, they offered me the job in Paradise.

My starting salary was $36,000/year and daycare was just over $400 a month. In the Bay Area I was making twice as much and paying four times as much for daycare. I rented the little blue house for just under $800/month and our only other expenses were food, clothes, gas, and my Beyond Fitness membership. It penciled just barely which was fine with me.

The Paradise Chamber had two other staff and many volunteers working in the visitor center. The volunteers had other pastimes and jobs including teaching square dancing, selling Mary Kay, and working at community events. Judy was the Chamber Treasurer and took care of the books, Melissa was the Board Chair, Kelley Connors was an Ambassador and business owner, and Pam Hartley was a member and owned Joy Lyn's Candies with her husband.

Right before leaving the Bay Area, I moved into a nice new office overlooking the Richmond Bridge and Mt. Tamalpais. At the Paradise Chamber, I looked out onto Skyway right across the street from Town Hall. A gas station, motel and pool, and diner were just down the street.

In the Bay Area, I commuted through the Caldecott Tunnel every day which, depending on which direction the third tunnel was operating, could take anywhere from 25 min to over an hour. In Paradise, my rental house was half a mile from my office going downhill, and a half a mile from Adela's preschool going uphill. My entire commute was under 4 min.

Shortly after I moved to Paradise, the garbage truck operator forgot to empty the Chamber's can. I shrugged it off thinking it'd have to wait another week. Nope. My office manager hopped right on the phone, "Tom? Yeah, Tom. Bill forgot to empty the Chamber's trash again, would you please send him back?" I was flabbergasted. If your garbage is forgotten in the Bay Area it stays forgotten, there was no Tom to call, nor Bill to send back. In Paradise, people operated on a first-name basis. Astonishing!

I started my job in Paradise in February of 2008. The Paradise Chamber hosts Johnny Appleseed Days which is famous for its volunteer-made apple pies. The prior Executive Director took me over to the Chamber's storage unit to show me the games for the Johnny Appleseed Days children's area. I will never forget how deflated I felt as that door raised revealing dusty, hand-painted plankboard games, apple-shaped bean bags, and carved wooden fish. Gone were the days of running expensive, international conventions with million-dollar budgets. Soon, though, Paradise taught me to see the value in small town life.

Where I grew up in the Bay Area, lawns are manicured, homes are well kept, and commercial areas are clean and tidy if not entirely brand new. I was used to a polished aesthetic I didn't realize at the time was not the norm throughout rural California. Paradise introduced me to another way of life. Before the Camp Fire, Paradise was a rare and special place where the community was built and tended by the people who lived there. Housing was not mass produced by developers with enough capital to stand up high rises. Rather, as soon as the paved roads veered away from the main routes, they turned to gravel.

I am not ashamed to say I did not understand Paradise when I arrived. More accurately and embarrassingly, my urban sensibility felt superior to rural life. Paradise was not filled with the sights and sounds I was used to: mass transit, international airports, 8-lane freeways, interstate junctions, and overcrossings. Paradise did not have critically-reviewed restaurants, global fashion outlets, nor internationally-renowned museums of modern art. Travelers could not catch a train or ferry from Paradise, nor fly to Singapore.

Gradually, though, the authenticity of the people in Paradise grabbed me and pulled me into the realest place I'd ever lived. My urban blinders came off to reveal the fact that having a homemade town was the sum total of having absolutely everything. Once I saw that, I realized two things: people who live in connection with their community have a worldliness you can't get from traveling the world, and though my urban upbringing is permanently affixed, my heart belongs to rural life.

Ultimately, I decided Chico with a population of just over 100,000 people is the right size city for me in Butte County.[11] I am close enough to everything I need for daily life, and not far from the places that make me feel whole, including big cities and itty-bitty communities. Fifteen years after leaving the Bay Area I'm part urban,

---

[11] Population count as of the 2020 US Census.

part rural, and part mixed up. I delight in the slower pace of life and when I get too intense, my husband tells me to "dial down the Bay Area."

I am endlessly grateful I got the chance to move outside of my urban comfort zone to experience Paradise before the Camp Fire. It was a town long after its time which made it very much before its time; a community time capsule that has a lot to teach us about quality of life.

Chapter 9
Community

In the spring of 2024, I attended an evacuation planning meeting at a charter school in a small intact community in the foothills that would soon be tested by the Park Fire. The meeting was squeezed between long volunteer shifts of brush clearing and tree trimming to make the school safer from wildfires. While the volunteers ate their lunches, our team shared the proposed evacuation routes and assembly points.

Hand-drawn signs were taped around the gym with safety notices and reminders about making eye contact with operators of heavy equipment before walking into their path. A large fox made entirely out of juice bottle caps was glued to the wall. As soon as the evacuation presentation was over, the air filled with the sound of tractors, diggers, fork lifts, chainsaws, and rakes as the volunteers loaded dumpsters with woody manzanita brush, drifts of pine needles, and ponderosa pine branches.

Most volunteers wore heavy work boots, hats, and safety vests and had their gloves tucked into their waistbands during lunch. One volunteer wearing thick leather chaps fired up a chainsaw just steps from the gym door after giving us his feedback. I watched our LA-based consultant watch the crowd. He and his assistant had flown into Sacramento and driven north to attend the meeting.

Understandably, the consultant focused on the maps printed on foam core standing on easels throughout the gym. These were the work product his firm was hired to create. He stood in front of the maps and puzzled over whether or not the only road visible on the maps should be the evacuation route bolded in red, or if all roads should be printed but in light gray. He pointed to the "not recommended" caption above an arrow on the map and asked if I thought the map should actually say that.

I told him my evacuation story from the Humboldt Fire in 2008, about how while everyone else was gridlocked heading up the hill I blasted down a road I wasn't sure was open. I told him I understand the goal is to point everyone toward the safest route, but not to the exclusion of all other routes if there is an obstruction or two. We talked about the disconnected road segments in Paradise that went from pavement to gravel to dead-ends halfway down the valley. I made sure he knew I was only

K. Simmons, *Three Fire Mountains*, https://doi.org/10.1007/978-3-032-17343-0_9

expressing my opinion from personal experience, and that the Sheriff's Office and CAL FIRE would know better.

Several times throughout the event I went outside to catch my breath from the heavy topic. Evacuation conversations are intense for me and they bring up strong feelings and memories as they do for just about everyone in Butte County. I take in a lot of trauma at work and am slowly learning over time how to care for myself in the process. While I stood alone in the sun, the campus hummed with crackling brush and trees, and tractors growling by, more soothing than talking about evacuation.

As much as out-of-town professionals bring to the table, I often wonder if locals bring more. In this case, they brought heavy manual labor to increase school safety, and lived experiences to our evacuation meeting. Importantly, residents know the location of each locked gate and which three people have a key. At the end of the meeting, our consultant told me he'd drive the "not recommended" routes so he could see first-hand what the Sheriff's Office meant when they advised against them for evacuation. We told him to look out for seasonal waterways, gates, poor gravel conditions, private bridges, dead and dying trees, one-lane roads, homemade warning signs, all the things we know might block or slow evacuation.

I was impressed by the volunteer work at the school site that day, and the fact that the group hung around to talk with us about evacuation. Without local knowledge and insight, evacuation maps might lack key information needed during an emergency. Our Sheriff's Deputy pointed out simple mistakes on the maps to me and said, "we can have five proofreaders, and it's never enough." Every single detail must be precise and accurate on evacuation maps.

Third-party consultants are necessary in recovery because they often come from outside of the area without baggage. They stay engaged in topics like evacuation planning without becoming emotionally entangled and bogged down by grief. It's fascinating to watch someone approach our greatest tragedy not by studying the community but by studying a computer model. There's tremendous value in the digital/paper approach, but if there's a reason the community balks at certain roads for evacuation, we need to know why. Our consultant gradually realized that studying a foam core map would never reveal the secrets of a one-lane, private road.

In much the same way I believe recovery can't be plotted on paper, neither can a perfect evacuation plan. It is essential to shake out each and every possibility imaginable, and it's necessary to remain on tippy toes for the inevitable scenario no one predicted. The Camp Fire threw everybody for a loop because of how quickly it arrived in Paradise from its ignition point, due to the direction and speed of the wind. With each fire, locals clock these details and re-set their evacuation plans according to the new information.

As residents, the environment teaches us as much as we are willing to learn and record with our human senses. It teaches us what the speed of the wind means, and the instability of mid-level moisture. It teaches us how to read variations in the arc and color of smoke plumes, the dangers of steep ridge fire acceleration, the risks of high fuel density, the value of managing vegetation, and the reliability of assembly points. When we are new to fire, we can sense some of these things in our bones. When we are attuned to fire, we keep our eyes to the sky and our ears to the wind.

I have learned to place my trust in people who know the land, who tend the land, who can tell the moment the wind isn't right. I have come to appreciate that mass public transit systems in urban areas are meant to service large populations anonymously, so life can be lived well beyond the scale of community. If every non-profit office manager called when the trash wasn't collected, systems might not keep pace with the required load. Urban flow exists to keep crowds in motion across large spaces. Rural life allows for individuals with lived experience to shape community.

9.1 Family Dinner

Around the same time we conducted evacuation outreach, we invited housing professionals from across the county to talk to Rural County Representatives of California staff who traveled from their offices in Sacramento to meet with us. The meeting honed in on programs designed to restore housing after disaster and the myriad challenges of implementation. The three-hour meeting progressed like a symphony, with everyone playing in perfect harmony. Locally, though we run many different programs, we share information and help each other through challenges.

Before the Camp Fire, jurisdictions in Butte County operated like an archipelago, many islands untouching in the sea. Cities have a distinct personality and character, and sit several miles apart. In the Bay Area, cities blend to the point you're never quite sure what city you're in unless you're somewhere in the middle. Los Angeles strikes me this way, with Imperial Highway moving seamlessly from city to city to city to city to city.

The demographics in each community were also distinct pre-fire, with affordability and median age differences being primary markers. Chico is home to Chico State which skews the population younger than Paradise which was known as a retirement community. The cities needed each other economically, but culturally functioned autonomously with little more than spheres of influence between them.

After the fire, the distance between the cities stayed the same but they grew much closer. Fire survivors found housing in Chico and Oroville. Gridley hosted one of the largest FEMA trailer sites.[1] Paradise employed the construction workforce. Fire has a way of revealing unseen truths, and what it revealed post-fire was a major interdependence between the cities for recovery. Those interdependencies aren't minor either, they are housing, infrastructure, recovery coordination, and healing.

At one point in the meeting with RCRC, as officials from each of the cities spoke around the table, I said, "this feels like a family dinner." I'm certain not everyone was in lock step, but they were committed to hanging together no matter what. The city administrator for one of the smaller cities said his community would be lost without County government.

The all-hands-on deck phase of Camp Fire recovery is over for the communities at large, but it is nowhere near over for the cities who are still navigating the same federal

---

[1] Observed through working in disaster recovery for Butte County 2021–2025.

funding regulations and State policies and procedures for recovery programs. We can finish each other's sentences when it comes to certain frustrations and obstacles to recovery.

At the meeting, the RCRC housing intern looked puzzled at one point. I told him I'd seen that face many times during disaster recovery. Federally-funded houses built to hurricane standards in California's wildland urban interface by a State contractor who is only familiar with flood recovery? Affordable housing projects scored to solve urban homelessness rather than rehousing rural families after disaster?[2] Recovery is puzzling.

Every person in that room was familiar with hitting their heads against the same walls over and over. We are powerless to the agencies who make the rules and enforce them. We explain, we show evidence, we provide case studies, we compare the number of homes built against the cost of administering the home-building program, we spot dangerous trends, we identify barriers, we articulate very clearly what we need in nice words and in firm words and sometimes in words of frustration. We do all the things we can think of to make a difference when we see a better way.

While recovery is a rocky landscape without a road, advocacy feels like jumping into a rushing river. When we've puzzled the advocates, we know we've described our circumstances well. They may never be as bitter or as angry as we are, but their arguments are made more powerful by rural wildfire recovery testimonials. Working with our statewide associations on issues prevalent across rural counties is the antidote to discouraging calls with our funders. In fact, calling an advocate and getting a sympathetic ear is sometimes the only relief for recovery rage.

9.2 System Outputs

At the same RCRC meeting, we talked about the design of each agency and funding source. W. Edwards Deming is attributed with saying, "The system is perfectly designed to get the results it gets."[3] If we're enraged by slow progress, we should look to the system that created it because that's a direct outcome of design—or lack thereof.

I have no formal training in systems thinking but I love it. On Career Day during my senior year of high school, we discovered the entire student body hadn't been assigned to groups to move from classroom to classroom to hear from professionals. I asked the office to print rosters of each class of kids—about 1,000 in all—and I had our Leadership Class take each roster and number the kids 1–30. We put the 1 s on a list, the 2 s on a list, and so on until we had thirty groups of kids which matched

---

[2] Observations and anecdotes provided by Seana O'Shaughnessy, Community Housing Improvement Program, 2024.

[3] Grow the Mind, Beyond Linear Thinking: Famous Systems Thinking Quotes, October 9, 2023. https://growthemind.ai/blogs/better-thinking/beyond-linear-thinking-famous-systems-thinking-quotes.

the number of speakers and rooms we had for rotation. (This was before laptops and excel.)

In our County Leadership Academy a few months back, about thirty of us were asked to guide a member of our group across fabric dots placed on the floor in rows of 8 by 8. There was a correct path known only by the facilitator who would "beep" when we stepped on an incorrect dot, requiring the stepper to backtrack correctly or get "beeped" again. We had to use trial and error to find the right dot sequence to get the stepper across without "beeps," and we had to rotate people with each failure until every stepper succeeded.

As soon as I understood the assignment, I suggested we assign four dots to eight line monitors on each side of the grid to watch and warn steppers about their dots. Sixteen people stepped up and so did the noise in the room.

The line monitors got really good at their jobs, pointing out the only dot per row that was safe to walk on. Another member enhanced the system by suggesting each line monitor have a back-up monitor so the original one could rotate across the dots, which gave us a way to get everyone through. Gradually, we learned our jobs, perfected our system, and in a matter of minutes got each person successfully across the dots without "beeps."

As much as I wish it did, disaster recovery doesn't work this way. First, there are no dots, no one really agrees on the objective, and plenty of people are "beeping." The room is crowded with steppers mis-stepping in all directions, and the "beeping" is out of control.

I learned early on in disaster recovery that there is no unified system keeping agencies, governments, and individuals in lock step through a coordinated, pre-planned, predictable process. Rather, it often seems there are competing systems operating within their own design, producing outputs that are misaligned at best and in conflict at worst.

Each funding source for disaster recovery has a different design. There are volumes of regulations that spell out every detail of each unique design, and different agencies that interact with and interpret each design. For example, CDBG-DR funding comes last by design. It is meant to be last-in money when all other sources have been exhausted, or when there is a gap no other funds can fill. If DR funds comprised 100% of the funding needed for multi-family housing projects, for example, they might quickly construct housing units that increase community viability and the pace of recovery.

Initially, Camp Fire CDBG-DR policies required that federal funds for multi-family housing construction not exceed 40% of the total project budget.[4] Other funding sources which are competitive, like tax credits and other subsidies, are required to complete the funding stack to make a project feasible for construction. If competitive rounds of funding don't align with the expenditure window for DR, or

---

[4] 2018 CDBG-DR Multifamily Housing Program Policies & Procedures Manual, Version 1.0, March 2020, CA Department of Housing and Community Development, page 12. https://www. hcd.ca.gov/community-development/disaster-recovery-programs/cdbg-dr/cdbg-dr-2017/docs/dr-mhp_pp_3-19-2020_508-compliant.pdf.

project scoring prioritizes different resident types within each funding source, then projects don't achieve full funding. Without full funding, housing units do not get built. As of late 2024, four of the County's nine multi-family housing projects were sitting in this position with significant funding gaps. If those gaps aren't filled, the County is at risk of losing the DR funds committed to the four unfunded projects, and the number of housing units those projects represent.

Because disaster recovery funding is so regulated, it's true that each agency and each appropriation is working perfectly as it was designed. In fact, each program is monitored to ensure that it is. The trick is not in understanding the design of each, it is in resolving the policy differences, misalignments, and conflicts between them. This is the hard work of disaster recovery. If DR prioritizes housing for displaced families, but State tax credits prioritize housing for elderly and disabled populations, then the very same project is going to score entirely differently on each scale. By design, the project is meant to score well on one side and poorly on the other because the funding sources are meant to meet different population needs. If these are the primary funding sources for multi-family housing post-disaster, and they don't tie together, then projects are not constructed.[5]

Is recovery designed to fail? Not intentionally, I hope, but I think it's fair to say recovery is not designed to succeed because it isn't designed at all. Legal action warned against a rural community almost entirely destroyed by fire is an output worth swimming back upstream to understand. There are a number of areas disaster recovery professionals can point to as evidence of design failures or design conflicts, and we document these in public comment letters on policy amendments. We can't make these changes at the local level, and we aren't very successful in moving the needle with the State, but we try anyway.

From my vantage point, State programs are meant to solve common urban problems, not uncommon rural recovery problems. If many visible social issues occur in urban areas where the density is, then solving those problems—or at least putting the money there—relieves the expectation of addressing rural issues. This could be true on its own, or due to the fact that rural concerns are less visible, misunderstood, and/or underprioritized for a variety of reasons.

I wonder how to design recovery if we can't agree on a basic definition of it. If the agencies managing funding meant to stack do not coordinate on environmental reviews, or if they communicate through a game of telephone with local government, or if they set expenditure periods that leave cracks large enough for conditional commitments to fall through, then recovery does not happen. In my experience, this a rule more than exception.

I believe there may be a way to design a system of recovery if we reconcile even these points. Developing a common language, agreed-upon terms, standard definitions, roles and responsibilities, system changes for close alignment particularly on the funding side—i.e., identical environmental reviews and eligibility requirements, with one approval process—could lead to a more cohesive, clarified pathway

---

[5] Fellowship Interview with Seana O'Shaughnessy, Community Housing Improvement Program, 2024.

for communities and individuals post-disaster. I recognize documents like FEMA's National Disaster Recovery Framework are meant to do this, but I believe only a system jointly designed by local, State, and federal agencies will fully account for various needs and obligations.

The Incident Command System (ICS) appears to function relatively seamlessly from the outside. Teams use phrases like "Type 1" to define a very specific standard of personnel and equipment. Common terminology encompasses relatively basic terms to describe highly technical distinctions.[6] Even the weather report is refined with specific phrases adapted to describe the impact of wind, heat, and moisture on fire. Witnessing these systems work feels like watching a clock tick: predictable and steady.

This is not to say everyone agrees on a single, correct, strategic approach to fire response—because I've seen disagreements as agencies work toward repopulation— but it is to say there is a system in place to shake out differences during incident management through the chain and span of command. This is not true in long-term recovery.

The phrase "too big to fail" comes to mind because I often think wildfire creates problems that are "too small to succeed" in rural communities where attention is on the disaster and not the aftermath and recovery. We know these challenges intimately, and we want to wave our arms in distress—and sometimes do to no avail—because the systems are national or statewide and are not designed to focus on us. I wonder how a single recovery system could be designed when disaster impacts are so varied, but if disasters themselves are so varied and humans can develop an incident command system to manage them, then surely we can figure this out, too.

There is not one single approach to recovery that everyone will agree is universally correct, but perhaps we can design a unified structure to hold space for working out differences as part of the process. I'm back to envisioning that table where all levels of governments sit, each with the proper authority to make decisions on behalf of their agencies, hashing out a design to support the process of recovery, while honoring the obligations on all sides.

In my early twenties I was handed the responsibility of negotiating multi-million-dollar contracts with international hotels for convention housing blocks. I learned during my very first phone call with a hotel in Toronto that starting with what we agreed upon led more easily into compromise on the things we didn't. I recommend the same approach for recovery system design, and would go so far as to guess we agree upon 90% of the pieces already, the rest gets us stuck.

If we had a recovery system designed to succeed, with places for hashing out disagreements, I bet disaster recovery professionals wouldn't feel so much rage. I bet legal action wouldn't be threatened, or if it was, the system would have a process for arbitrating differences before they escalate. It feels like a giant waste of energy to lament the comprehensive failures of recovery when the individual systems are designed to produce exactly these results. On the same token, if we see cracks in various recovery processes and we don't advocate for change, then we allow our

---

[6] FEMA ICS-100 class required by Butte County for disaster workers.

communities to suffer silently by design. For this reason, Butte County is full of advocates.

I walked into Stanford University thinking we are recovering too slowly. Now I see value in digging deeper into the conflicting policies and priorities agencies use to fund and pace our recovery. It's not for lack of effort or creativity on the local side, nor on the State and federal side. The disconnect might be mostly attributable to lack of familiarity with local conditions and the inevitable degrees of separation caused by geography and span of responsibility.

Recently, I watched a State employee realize that by listing the Camp Fire's disaster declaration, "DR-4407," on a program flyer open to everyone, non-impacted residents would not apply. Federal disaster jargon that typically relates only to fire survivors decreases program uptake by non-impacted residents who haven't spent years learning the language. In rare cases where State programs are open to everyone in the county—like housing mitigation grants—materials should be written in common language without reference to the disaster.

The County participates in Emergency Operations Center trainings for staff who serve in the EOC if disaster strikes. These trainings emphasize the use of plain language when communicating during disaster. Plain language is important for the conveyance of critical information when call signs and radio codes are different even between cities within the same county. I think about the need for plain language in the disaster recovery field. Complicated acronyms for grants like FMAG, CDAA, HMGP, CDBG-DR-MHP, CDBG-DR-MIT-PPS, can be very overwhelming to a brand-new workforce who shows up to help. Even knowing what these mean and using them daily myself, I want to swat them away like flies. I joke in meetings with non-recovery staff that I fall asleep when I hear myself speak in acronyms. I'm as bad as the FEMA manuals because I have to be.

I receive emails from the State describing procedures for reporting on grant expenses and see that while they are technically correct, they are speaking an entirely different language than we are. When onboarding disaster recovery staff, I spend a lot of time in front of a white board drawing lines between agencies to show the flow of funds, and spell out acronyms so they know what funding comes from which agency. A representative from a State agency told me it takes an average of four years to master federal regulations for disaster recovery grant administration, which is a long time to suspend confidence on the job. Until staff speaks the language, it can be intimidating to converse in a meaningful way, let alone ask questions.

To demonstrate readiness for federal grant administration, local staff are required to submit organizational charts and resumes to the State as due diligence. The State is looking for acumen in administering federal disaster grants which is fair, but where does that skill set come from if not from a prior disaster? In a jurisdiction where disaster strikes once, say, staff will be entirely new to disaster grant administration from day one of recovery. In a community like Butte where we are dealing with multiple disasters, we certainly have staff with some experience, but the State appears to be looking for years, perhaps decades, of time spent administering these funds, when they are highly specialized and specific to one-time incidents. Thankfully, the State offers what they call Technical Assistance (TA) to guide staff through the grant administration process. Even grasping the concept of TA took me a beat. "Grant Coaching" might be a better way to describe TA in plain language.

On a coordinating call for the owner-occupied reconstruction program for North Complex survivors, our Development Services Director told HCD staff that their zoom backgrounds with neat suburban neighborhoods do not mirror our reality. I can't say for sure that State housing programs work better in urban areas than they do in rural areas because I've only tried to make them work in rural areas. But RCRC staff did tell us during our housing roundtable that urban areas receive a disproportionally higher amount of housing tax credits than rural areas.

A few years into recovery from the 2018 fires, the State is proposing to shift funding from the Owner-Occupied Reconstruction (OOR) Program to Homebuyer Assistance and Multi-Family Housing Programs due to a lack of OOR applicants. The Homebuyer Assistance Program aims to relocate survivors into lower-risk areas for wildfire, into what I call "displacement housing." The State's initial goal was to assist survivors in recovering in place, allowing them to remain on their properties and in their communities in permanent housing. For a variety of reasons, survivors are reluctant to accept loans or grants from the government, or they are ineligible for the funding due to spending their insurance proceeds on RVs soon after the fire. Unable to "reimburse" the State for the cost of their RV to qualify for a federal loan, these survivors are left without permanent housing.

Like the movement toward aging in place, wildfire recovery should aim for recovering in place. Converting owners to renters by adding multi-family housing in urban areas rather than single family homes in the burn scar, and incentivizing relocation to lower risk areas, cuts into a community's ability to recover in place. Certainly, the cost of home ownership and maintaining insurance is too burdensome for many survivors post-fire, but government programs should adapt their criteria to what survivors need to thrive in place.

I've always suspected that the local hand moving the chess piece on the game board of recovery isn't working independently of the State and federal governments. We are constantly questioning and interpreting federal and State guidance to make sense of what's written, let alone what isn't written. Are there too many disasters of different kinds to expect a tight level of coordination between government agencies? Without a system designed for success in place, yes. Are too many disaster-stricken communities flailing through recovery without a lack of system and design? Tragically, yes.

Equity, rationality, adaptation—everything we're after with recovery—lie somewhere between possibility and perfection. If we know communities and survivors are being left behind by programs that are abjectly failing in rural areas because they're designed to succeed elsewhere, then we are either accepting a permanent loss of our rural way of life, or we are enraged and calling our lobbyists. Neither option is sustainable, and neither option leads to recovery.

9.3 On Paper

In the foothills, many communities are served by one road in and out which makes agreeing on the primary evacuation route fairly straightforward. It's when we get to the dirt roads that words begin to matter. Initially, our evacuation planning team preferred, "not a recommended evacuation route," for the roads that have one or more impediments like lack of regular maintenance. This suggests the road can be used for evacuation but isn't a recommended route versus, "not an evacuation route," which signals the road is impassable in an emergency. One word might be the difference between life and death.

My mom escaped the Camp Fire by driving out of Butte Creek Canyon on a bike path. Currently, our evacuation maps do not list bike paths as recommended evacuation routes, but we know they're used out of necessity. In fact, funders are becoming attuned to transportation projects that incorporate bike paths for additional evacuation. Paper planning is a great conversation starter and necessary for broad public messaging, but can't possibly encompass all variables.

Listening to our Sheriff talk about Camp Fire evacuation, it is clear every feasible option is considered by law enforcement out in the field given the circumstances. With each passing moment, fire challenges, limits, and compromises those options. In mitigation planning, there is great responsibility for documenting generalized expectations for the public so they have a place to start, but during emergencies the responsibility lies with first responders making decisions as quickly and accurately as they can. Residents in high-risk foothills should study the conditions of their surrounding areas as well as any map.

When I worked in recovery for the Town I was contacted by a volunteer in a small, second-home community in the mountains above the Sacramento valley. Many of the second homes are owned by people primarily working and living in the Bay Area. This foothill community is served by one form of ingress and egress bisected by railroad tracks.

The volunteer heard the Town of Paradise was looking at installing an early warning siren system. The siren system was identified as a priority project in the Long-Term Community Recovery Plan following the Camp Fire.[7] During the Town's feasibility study for the siren system, we conducted a community survey and I read through each of the 1,100 comments. The survey indicated a strong, favorable response for installing siren towers among current and displaced Paradise residents, referencing both the full communication systems failures during the Camp Fire and the speed at which the fire moved. Every comment bore a survivor's story.

I was invited to present the Town's project at a virtual fundraiser for the installation of a siren tower in this Sacramento foothill community. My presentation covered the basics of the Town pre-fire, the Camp Fire maps, the recommendations of the After Action Report to address communications systems failures, and general community

---

[7] Town of Paradise Long-Term Community Recovery Plan, June 2019, Urban Design Associates. https://www.townofparadise.com/sites/default/files/fileattachments/recovery/page/42792/062519_final_recovery_plan_compressed.pdf.

perceptions about siren tower installation. Afterward, the volunteer told me she raised over \$20,000 in that one event for the single siren tower they needed. As a next step, I recommended a professional feasibility analysis, and that was the last we spoke.

It's fair to say disaster recovery is driven by the search for a permanent fix for fear. What is the one project, the one program, the one map we can create that will reduce risk and save lives. In practice, there should be layers upon layers of redundancy, duplication, back-up, back-ups for the back-up, and whatever else works and is affordable. I could hear in the volunteer's voice that she believed it was only a matter of time before her community burned as well.

I grew up on a cul-de-sac in Moraga and never once thought about wildfire evacuation. It never occurred to me, and possibly to my parents, that we might need to make a plan to get out. We established a meeting point outside of our house because that's what we were taught in school, but we never talked about "go bags" or an exit route from the neighborhood. Even when I got my driver's license, I didn't think about whether I would escape through Orinda or Lafayette, or the Oakland Hills if I needed to. Moraga does not sit on a major freeway like many Bay Area suburbs, so getting out would always require going through somewhere else.

We bought our corner house in Chico one street over from the freeway onramp a few years before the Camp Fire. I've thought hundreds, maybe thousands, of times about how glad I am that we live on a corner and can go at least three different directions to escape. I live in a constant state of readiness, and wonder what that'll do to my peace of mind if it never goes away.

A few blocks outside of our neighborhood is a relatively new development with homes built on one side of the street due to the narrow roadway. Residents park their cars on the sidewalk because parking at the curb would block the narrow lane. Walking down this street I get chills thinking about everyone trying to evacuate at once; cars would not be able to maneuver beyond their driveways.

I do not recommend living in fear, but I do recommend being prepared, and that's what evacuation maps are for. County stakeholders review decisions made for old maps and fold in real evacuation experiences from the Oroville Spillway Incident which forced a mass evacuation of 200,000 people, the Camp Fire, the North Complex Fire, the Park Fire, all fires in the past and the fires we think about for the future.

9.4 Long Goodbye

The funny thing about grief is you grieve the gradual fading of grief itself, as the thing you've lost slips further away. In the first years after my dad died, I'd wake up in the morning surprised to remember he was gone. Every once in a while, I still feel compelled to call him when I have a great story. For the most part, though, I know he's gone and that loss is now as much a part of my life as his life once was.

Losing Paradise felt similar. In the days after the Camp Fire, I'd wake up in the morning shocked to remember Paradise was gone. There was no gradual acceptance over time, just an awful daily realization that struck me in my bones each morning.

I'd feel the full weight of it on my skin when I thought about the people affected, including my parents, the pressure of sudden loss tightening like a vice.

Learning about my nerve disease, I know there's a physiological reason my feelings show up on my skin. The myelin sheath covering my entire peripheral nervous system is unraveling leaving my body vulnerable to sensations like heat, cold, and pain. When Adela was in a car-jacking at nine years old, my skin felt like it was on fire for a whole year—not a faint sensory experience, but burning an inch deep.

Stress shows up in my ears, too. If you've ever heard the distant sound of a rock slide while camping in Yosemite, stress sounds like that: a gradual, growing rumble, hinting of danger. Commuting with debris and tree removal trucks, I hear this rock slide in my ears with the windows up and the music on.

Thankfully, the pain on my skin and rumbling in my ears is easing up. It's becoming something that starts and stops on occasion, rather than something I experience chronically. I credit this partly to the passage of time and also to the fact I don't want to be on this emotional journey anymore. My nerves need a rest.

The beauty of working on recovery long term is a gradual distancing from the trauma response, and the ability to see it for what it is so I can make choices that protect me. I know which calls, meetings, and experiences are going to upset me before they happen now, and I either prepare myself or minimize their duration. Though I can't totally turn off my trauma response, I know how to live in the thick of it for short periods of time, which makes it easier to bear. I wish I had a calm, cool, unrufflable demeanor but my hands shake, my voice quakes, and I have less patience when I'm knee deep, but I know why it's happening so I give myself grace.

I've come across plenty of people who exhibit a trauma response when we talk about our fires. Depending on the topic and the time we have together, it can last a moment or a whole meeting. I also watch consultants grapple with the trauma response they get from their wildfire-impacted clients. Like most people, they lack the language and framework for acknowledging it. I hope for communities impacted by wildfire in the future, trauma is a better understood, cared for silent partner in recovery, rather than the elephant everyone stares at in the corner.

Similar to how the visceral memory of my dad is fading, so is my memory of Paradise before the fire. Now when I'm on the ridge, I don't expect to see anything but what it looks like now, and with the fire debris slowly coming out, I have less reference for what it used to be. Rather than the skeletal trees reminding me of the great forests, now there are fields and stumps and wildflowers and vistas. Shortly after we sold my dad's house, I drove by half expecting to see him tinkering under the hood of his car in the driveway. Now when we drive by it looks like somebody else's house, with a few markers of the past like the tree he planted and the sidewalk stone he puzzle-pieced together.

I feel resistance to letting go of my grief for Paradise because I'm afraid to lose my memories completely. The grief keeps me tethered to the life we had for that brief moment in time: Adela's youth, my confidence. But I'm exhausted by the buzzing on my skin, the shortness of my breath, and the sound of sadness filling my ears. I don't want to live at the mercy of remembering a place that isn't here anymore when it is

NTP	Disaster Infrastructure	SCOR Plant Upgrade	CDBG-DR (18-DRINFRA-18001-00020)
Application	Disaster Infrastructure	Facility - Gridley	CDBG-DR Infrastructure
Application	Disaster Infrastructure	Facility - Paradise	CDBG-DR Infrastructure
Application	Disaster Infrastructure	Facility - Oroville	CDBG-DR Infrastructure
Application	Disaster Infrastructure	Road Reconstruction	CDBG-DR Infrastructure
NOI	Disaster Infrastructure	Road Reconstruction (FEMA PA Match Only)	CDBG-DR Infrastructure
Awarded	Disaster Mitigation	Evacuation Planning	CDBG-DR (17-MITPPS-21001)
Awarded	Disaster Mitigation	Code Enforcement - 38A	CDBG-DR (17-MITPPS-21002)
Awarded	Disaster Mitigation	Fire Protection & Prevention Education	CDBG-DR (17-MITPPS-21003)
Application	Disaster Mitigation	Broadband Planning	CDBG-DR MIT-PPS
Application	Disaster Mitigation	Fire Protection & Prevention Outreach/Education Expansion	CDBG-DR MIT-PPS
Application	Disaster Mitigation	Fire Protection & Prevention Code Enforcement Expansion	CDBG-DR MIT-PPS
Application	Disaster Mitigation	Foothill Rebuild Barrier Removal	CDBG-DR MIT-PPS
Application	Disaster Mitigation	Older Adult Housing Support	CDBG-DR MIT-PPS
Application	Disaster Mitigation	Rural Water Safety	CDBG-DR MIT-PPS
Application	Disaster Mitigation	SERVE Access Function Needs Training	CDBG-DR MIT-PPS
Application	Disaster Mitigation	EOC Capacity Planning	CDBG-DR MIT-PPS
Application	Disaster Mitigation	Emergency Operations Plan	CDBG-DR MIT-PPS
Application	Disaster Mitigation	Roadside Fuel Reduction Plan	CDBG-DR MIT-PPS

Fig. 9.1 Chart showing various disaster recovery grants managed by Butte County staff

changing every day. Fire, aftermath, recovery—none of these are static conditions, nor should my mental state be.

The last time I onboarded new staff, I created a giant spreadsheet of all the grants we manage in our division at work. Out of more than 50, all but eight are related to federally declared disasters: 2017 Wind Complex, 2018 Camp Fire, and 2020 North Complex Fire. A handful are related to COVID but those are phasing out, whereas Butte County will be monitoring disaster-funded multi-family housing projects for several decades. Some of our "tasks" in the State grant portal are due in 2051. So, while I work my way into an emotional place where the edges are less sharp from both the fire and the recovery, I can't leave either behind entirely. Signing up to work in recovery means signing up to live with disaster for a very long time (Fig. 9.1).

Personal resilience isn't a foregone conclusion after disaster. Wildfire creates wreckage, and resilience doesn't magically appear at the wave of a recovery wand. The more I search for resilience in myself and within the recovery field, the more I see it as wholly separate from both. Resilience is a choice to interfere with the natural progression of things by using tools that lead to outcomes that do not match the past.

For me, achieving a resilient mindset means having more curiosity for the joy and triumph in life after recovery, than for the damage and devastation inseparable from the process. The slog of long recovery does not offer much room to explore joy and triumph when community reconstruction moves at a rate of 3% per year, so interference is necessary. Resilience requires saying goodbye to grief so I can embrace a future that looks nothing like the past.

9.5 River Crossing

I imagine someone reading this might wonder why anyone would spend five years grieving a town. If I hadn't witnessed the Camp Fire, I might wonder, too, because while our small community has been grieving, billions of people are living their daily lives in intact towns, cities, metropolises, crossing bridges, riding elevators, walking to the park thinking nothing about it. I remember living like this before I knew whole towns could disappear.

In March of 2024, a cargo ship hit a bridge in Baltimore and it collapsed into the river and onto the bow of the ship.[8] In California, this meant round-the-clock news coverage. The bridge collapsed over and over on television screens lining our local gym's walls for several days. I heard a newscaster say in disbelief that the city would be recovering from this incident for several years, as if reading a script she didn't quite believe.

Instantly after the collapse, the media reported elected officials calling for a congressional appropriation of federal funds to fully reconstruct the bridge. I am not close enough to the event nor do I watch the news enough to know if this appropriation was granted, but I am certain the bridge will be rebuilt for how much the city relies upon it. The economic impact of the bridge collapse was immediate: the port closed temporarily and a major connection between two points in the city was gone. In recovery, the city will find new avenues for commerce and traffic, but the lives lost are irrecoverable.

I can't watch images of the Twin Towers falling on September 11th, videos of the Camp Fire progressing, nor news footage of the bridge falling with the twinkling lights of construction work on the closest span to the ship. It's amazing we don't notice the harm repetitive disaster imagery does to our consciousness, even a looping news clip. I don't have whatever toughness it takes to absorb these tragedies on repeat, perhaps because I know the long walk after disaster.

In my former life, I commuted over bridges in the Bay Area daily. My first thought upon seeing the Baltimore bridge collapse was: where will this happen next? It's hard to accept something like this as an isolated incident when cargo ships pass under bridges all over the world every day. Major bridges collapse so infrequently they make national news, yet even one incident makes us wonder about the likelihood it will happen again. We question this after disaster because what we really want to know is: am I safe?

When the media brings us images of tragedy befalling another community, we tend to tighten the bolts and screws under our own feet. I'm fairly certain agencies across the country felt differently about the bridges under their purview the morning after the collapse. I imagine crews inspecting each span to identify weaknesses they may not have seen nor thought to look for before. Sometimes, infrastructure shows warning signs of imminent catastrophic failure like water coursing over the emergency spillway at the Oroville Dam. But, given the infinite circumstances infrastructure is built to withstand, these failures are sometimes impossible to predict. Perhaps seeing a national disaster shakes the dust off of we don't want to think about but must.

Like the cliche, "hug your kids tighter," after tragedy, we should check our bridges, our dams, and our buildings for the worst that can happen We should turn fresh eyes on overgrown brush after fire. I'm sure thousands of people working in high rises wondered about the ability of their buildings to withstand airplane attacks after 9–11.

---

[8] Hutchinson (2024).

And today, ship crews are probably measuring the height of their cargo differently and perhaps testing their back-up power systems more frequently. These incidents are good reminders to mitigate risk.

After the 1989 earthquake, I regularly imagined another "Big One" happening while I was crossing the Bay Bridge. The new span is gorgeous with steel bars fanning across like eastern portion like permanent outriggers, perhaps designed to create confidence in human engineering following the quake. I can imagine those design meetings post-quake, the team acknowledging their responsibility to instill confidence in generations to come with a strong aesthetic. At least that's what I hope for: someone recognizing how generations might feel crossing the new bridge after witnessing the last bridge fail.

If the agencies in charge of reconstructing the Baltimore bridge use federal funds, it will take years. Federal funds are neither fast, free, nor flexible. Any liability payments and settlements for damages will have to be assessed for duplication of benefits. Aside from the funding, the public will expect the government to replace the bridge quickly. In our private lives after disaster, lack of sufficient funding is often the primary barrier to individual recovery. In the public sector, access to funding can be the greatest complication for recovery, prohibitively increasing the local cost-share by requiring compliance with federal regulations.

I imagine some Baltimore residents may call for an exact replica of the historic bridge design, others will call for resilient infrastructure no matter the design. There are agencies operating and maintaining the bridge on paper, and the general public who doesn't know the skyline without it. Looking across the river without the bridge is probably as disorienting as the absence of the Twin Towers on the New York skyline, as disorienting as looking for my rental house after the Camp Fire.

I hope the public and various levels of government in Baltimore come together to rebuild the bridge with a unified plan. I hope solicitations flown for construction using federal disaster grants are not contested. I hope materials and labor are readily available and provide opportunities to the local workforce. I hope the design finds middle ground between historic and resilient if they're at odds, and the new bridge becomes something the city can embrace as a healing, iconic fixture. I hope those on the cargo ship and rescue teams seek and receive help for their experiences. I hope the families of those who passed find peace.

Finally, when the bridge is fully reconstructed, I hope it feels like an accomplishment. I hope workers transition to other jobs seamlessly, and this project doesn't overshadow the next. I hope the community is able to heal by crossing the bridge together for the first time, trusting the solid ground beneath their feet. I hope recovery isn't something they have to recover from.

Reference

Hutchinson B (March 28, 2024) Baltimore bridge collapse timeline: inside the cargo ship collision. ABC News. https://abcnews.go.com/US/timeline-baltimores-key-bridge-collapse-shows-moments-cargo/story?id=108540377

Chapter 10
Mending

One Sunday in October of 2016, my sister called and told me she was pregnant with her first child. At the time, she lived with my dad in Petaluma and as I paced my backyard in Chico listening to her, I couldn't be happier. That Thursday evening my dad called with news of his own. With another grandbaby on the way he decided it was finally time to get sober. After many years of fits and starts with sobriety, he was following the advice of a friend and taking himself to the Emergency Room that night.

I took the next day off of work to drive down and help my sister help my dad. At the Emergency Room the night before, he was given a prescription for detoxing at home and a referral for an in-patient treatment center. Because it was the end of the week, he'd report to the treatment center on Monday morning for an intake consultation. He was released from the ER and sent home for the weekend.

When I arrived in Petaluma that Friday morning, I loaded him up and drove to the hospital pharmacy in Santa Rosa where we picked up his prescription. It had been a few weeks since I'd seen him but I was surprised by the way he leaned his full body onto the car and talked nonsense to me, looking me straight in the eyes. His texts leading up to that day had turned into single letters and numbers meant to convey brief pep talks. I remember sitting in a work meeting and hearing my phone buzz 16 times as each letter arrived individually. His very last text to me was "Luvu2."

The pharmacist that day was specific about the rules of home-based detox but assured us we'd be fine. We took the pills back to his house, separated them into daily doses according to the instructions, and dad was set for the weekend. Looking back on this now, I remember feeling that same push–pull I did when driving through fire about dad detoxing at home; one foot on the gas pedal, one foot on the brake.

Later that afternoon, consistent with his declining behavior, dad began hallucinating from his bed, talking loudly to people who weren't there. I'd later learn from a sobriety professional that these were lucid dreams which is the only sleep he got toward the end. We did not feel confident in our ability to care for him so we took him back to the Emergency Room that night. When the doctor came to talk with us,

K. Simmons, *Three Fire Mountains*, https://doi.org/10.1007/978-3-032-17343-0_10

he asked my dad how much alcohol he consumed each day. My dad made a couple of jumbly jokes then said two quarts of vodka, before pausing and grunting it was usually three. Three quarts of vodka? Dad! I was shocked.

The ER staff admitted him for observation because he was neither coherent nor able to get himself around very easily. I sat next to his hospital bed and watched him sleep behind the oxygen mask feeling a wary sense of relief. I hoped they'd keep him in the hospital until Monday morning when we could get him into the treatment center.

Feeling like I could exhale briefly, I went for a walk around the hospital grounds and watched the sunset. I called my uncle, my dad's younger brother, who was home in Santa Cruz with a stomach bug and told him what I'd learned about dad's alcohol consumption. When I got back to the waiting room there was dad, sitting alone in a chair wearing gray sweatpants and a shirt, hands on his knees, grinning like he was waiting to be picked up from school.

I was flabbergasted. I walked over to the nurse's station and asked why he'd been released. They told me they'd checked him out because there was nothing they could do for him. "But…" I stammered, "he can't sleep without the oxygen mask, can't he at least stay here for that!?" They told me no. When he stood up to walk with me to the car, I saw he'd urinated down the front of his clothes. I wanted to turn around and scream, "do you see what is happening here?!".

In the car ride home, he gazed out the passenger window and ordered a pizza into the cool night air. He asked for extra cheese, extra sauce, and mushrooms on thick crust, pausing now and again for his signature chuckle. My sister's boyfriend who was driving the car looked at me in the rearview mirror and we shook our heads. Dad had been hallucinating for months because of the alcoholism so this wasn't new, but we were not equipped to handle it.

Once we got dad home and back into bed where he continued to call out various names, my sister and I agreed on a short-term plan for the round-the-clock care he clearly needed. She'd be there that night while I drove home to pack for the week, then I'd be back the next day. That night in bed, my husband held my hand as if to say, "we can do this." During our annual family camping trip to Lake Pillsbury just four months earlier, I said I didn't think dad would live past the end of the year. But here I was, hovering in that rare sliver of hope children of addiction allow themselves to visit when there's good reason to believe change is coming. He had an intake form for a treatment center, we just had to make it through two more days before he was admitted and on the road to recovery.

The next morning while pacing my driveway I called the treatment center to see if they could get him in earlier. They told me they couldn't admit him on the weekend, but once I described the hallucinations, the visions, the weeks of sleeplessness, the physical downward spiral, they told me he was nearing the terminal stage of alcoholism. They assured me they'd be there to help when we saw them on Monday.

I hopped back in the car toward Petaluma and was an hour down the road when my sister called, frantic and hysterical. She was at the neighbor's house because dad's legs were kicking violently and she'd called an ambulance. I tried my best to calm her down while concentrating on the only curvy section of I-5 heading south through

Arbuckle. I told her he would be ok, help had arrived, they'd take him to the hospital where he needed to be.

My sister fell quiet as she listened to another voice in the room. Suddenly she was screaming, "He's dead! He's dead!" over and over. Hyperventilating now, I pulled my car over while I heard the paramedics, her boyfriend, the neighbors move in to console her. My world was spinning.

When I was able to get back on the interstate, I called my uncle who was now getting in the car himself, sick or not. I didn't realize I'd missed my turnoff for the Bay Area until I reached Sacramento, and even then, had to pull off the freeway to figure out east from west.

Short lived as it was, my dad died with recovery in his heart which has come to feel like a gift. Only when I sat with him and the Chaplain a few hours later did I feel the powerful absence of his addiction. It left him when he left us, and I know he is finally at peace.

Mendocino Complex Fire, July 27, 2018—January 4, 2019

When my dad was a young boy, he and his family spent summers in Lake Pillsbury. Coincidentally, my mom and her family did, too. She is the second of five children born within six years. My sister and I arrived 17 months apart and my parents carried on the tradition of going to Lake Pillsbury for long weekends in the summer. Even after they divorced, they'd still go up to Lake Pillsbury, my mom for one week in June, my dad for two weeks in in August, gradually aligning their vacations until we were all up there at the same time.

My 20-year-old daughter was six months old during her first summer trip to Pillsbury; my 12-year-old was almost one which means the year prior I braved the heat and dirt nearly nine months pregnant. Pillsbury is not an easy place to take kids but it is the best place in the world for kids. To this day, both sides of my extended family go to Pillsbury each year: all of my mom's siblings, their spouses, their kids, their kids' kids, even my dad's brother comes up on occasion. My grandparents would make the trip when they were still alive, walking down the rocky path to sit and watch the kids swim from the Liar's Bench. My mom's oldest brother and his sons fly across the country to drive that rutty dirt road to our family's version of paradise.

Lake Pillsbury is a man-made lake behind a hydro-electric dam owned by PG&E in Lake County, California. Boaters tow skiers, tubers, knee boarders, bare-footers, anything a boat will tow. My Uncle Tim loves to putter his tiny silver fishing boat out onto the lake every morning when the mist is inches off the surface. My Uncle Tom bought my dad's ski boat after he died, and captains the tube and wakeboard crews now, taking the torch from Uncle Matt who doesn't bring his boat up anymore.

From toddlerhood through my early teens, I spent the first two weeks in August in Lake Pillsbury with my dad, then we all shifted to the last week in June to match up with my mom's family who'd struck up their childhood tradition again. Dad would load his boat to the brim with camping supplies and frozen barbecue meat, then strap the whole thing down under a tarp so the dirt wouldn't cake onto our tents, sleeping bags, and food. Once the truck was unloaded and the boat was anchored in the lake, my sister and I loved to sit in the back of his pick-up as we bounced and fishtailed over miles of gravel road around the lake. Pillsbury was my dad's happy place.

I had crushes at the lake, brought my friends sometimes, eventually boyfriends, my husband, and my daughters. I grew up at the lake. Heidi recently pulled out the Pillsbury list to refresh it before we start packing. I watched her erase something and told her the list doesn't have to be perfect because Pillsbury isn't perfect. She said, "Mom, Pillsbury is the most perfect place there is."

The Mendocino Complex Fire started one month after our family returned home from the lake in 2018.[1] I'd sold my dad's house in 2017 during the Tubbs Fire in Santa Rosa which halted new insurance policies in Petaluma for a time, delaying the close of escrow. Now, we watched the news in concern then alarm as this new fire moved closer to Lake Pillsbury. We'd seen a handful of close calls over the years, but this fire came on the heels of the Tubbs Fire and struck real fear. The era of the mega fire had begun.

My parents, siblings, uncles, aunts, cousins—the whole crew—starting calling and texting each other as the fire grew. We watched the fire map push up toward the lake through the forests we'd driven through for decades. The fire widened and lengthened almost as if reaching for Lake Pillsbury which it eventually did, touching down near the Rice Fork on the back side of the dam.

There's a point when a fire is burning close to everything you love that you begin bargaining. Take the forests, the hills, the mountains, we thought selfishly, but leave the cabins, the general store, the old wooden trout fishing signs, the rope swings, the Liar's Bench. The owners of the cabins and marina posted updates of firefighters camping in our spots, tired, dirty, worn. The fire ringed a portion of the lake but left the cabins and our precious memories intact. Our bargaining had worked but we had yet to take in the full extent of the destruction, realizing once we did that Lake Pillsbury isn't the same without its wilderness.

And while we held our breaths for Lake Pillsbury, the Camp Fire destroyed Paradise.

All I could think while portions of our favorite place in the world were burning was how shocked and devastated my dad would be. Equal to our grief for him would be his grief for that lake. When we poured his ashes into the Rice Fork from the back of his boat in 2022, finally ready to let him swim with the fishes, we did so amidst a backdrop of blackened trees.

There's grief for places lost to fire and grief for places that are bound to burn. Lake Pillsbury is one of those places for me. It is home to the happiest childhood memories of generations of my family, the burial site of my dad, the place we all want our ashes spread, too, and the center of my soul. Every year now, in this season of fire after fire after overlapping, layering, suffocating, life-erasing fire, I say goodbye to Lake Pillsbury knowing it may be the last time I see it before it burns. PG&E is decommissioning the dam and selling off many of their hydroelectric facilities,

---

[1] CAL Fire Incident Report, Ranch Fire (Mendocino Complex), last updated 10/24/2022, https://www.fire.ca.gov/incidents/2018/7/27/ranch-fire-mendocino-complex.

likely due to the settlements and lawsuits stemming from utility-caused wildfires.[2] Their plans to remove the dam don't break my heart like the thought of fire coming for Pillsbury.

Someday in the not-too-distant future Lake Pillsbury will likely be drained. The rivershed will slowly adapt its shape and size to a new volume of water flowing through. Wildlife will adjust to the changes in their wet lands, and the elk herd may travel from their flooded meadow under Hull Mountain to another site up-river for food. Any homes that derive their value from the lake front may be left unoccupied until the viewshed is restored, or those families will adapt like ours is adapting now.

If the lake returns to a river, our little resort will be left high and dry. The owners who operate the business may end their lease with the national forest. The cabins and bathrooms may stand for a while, eventually crumbling, or they may be kept up by a proprietor with a different vision. We don't know. As sad as this unknown is, I like to think the trees, brush, and wildlife that lived in the watershed before the lake was created may prefer the river anyhow. Time may transform our favorite area into something unrecognizable to us but very familiar to nature. This is what I say to my mom every time she cries about the lake.

The thought of losing Pillsbury to fire, however, is unbearable. It's a threat of erasing not only the fun we've experienced but the tie that binds our family together through divorce, illness, death, marriage, childbirth, and everything in between. For me, it's the alternate universe in which my parents co-existed in peace long after their differences drove them apart in the real world. Adela's dad and his girlfriend camp right across the way from me and my husband, and we eat dinner together every night. Pillsbury is a magic glue more powerful than any division humans create between themselves, and offers itself up as the only place these memories can be made.

We've talked about finding another spot to spend time together each summer, but so far these conversations have not materialized into any plans. Nobody wants to let go of Lake Pillsbury until Lake Pillsbury lets go of us.

For now, we will continue refining our list in April, start the piling and packing in May, then caravan up to the lake in June. If fire gets there first, we'll have memories that last our lifetimes and the lifetimes of those who came before us that now settle as sparkles at the bottom of the lake.

Fire or not, Lake Pillsbury is changing right before our very eyes, disappearing one way or another. I pledge to love it no matter what, just as it has always loved us.

---

[2] Patti Poppe, PG&E CEO, Presentation at Rural Representatives of California, October 2025 Board Meeting, Olympic Valley, confirmed decommissioning.

10.1   Recovery Reluctance

In April of 2024, I attended a celebration honoring the bravery of victims of crime in Butte County who'd been assisted by the District Attorney's Victim Assistance Bureau.[3] As each survivor was honored, the speakers mentioned their personal recovery as an essential part of their ability to thrive today. For some, the recovery was physical from a car accident or attack. For others, recovery was emotional and required deep perseverance. Each person offered a recovery story as evidence of their healing and the power of support.

As I listened to the speakers I wondered about my own recovery, and why I hadn't taken it seriously or made a plan. Maybe because there wasn't a single incident I could point to that required recovery, just days piling on days living the aftermath of wildfire and disaster recovery. I'd dabbled in recovery activities but never for the purpose of personal recovery.

I attended another resilience training recently and listened again to the necessity of recovering from trauma. The trainers described a process of moving from a "before" state through an incident, from suffering into coping, then past a turning point into thriving, and eventually growing beyond the original state. I could feel the invitation to recover and paid attention to my reluctance, and to the compounding but blurry effects of my years-long experience with daily hardship versus a single traumatic event.

The act of recovery requires trusting in a future that will be different from the past—maybe even better. I believe this is what my dad sought when we learned he'd soon have another grandchild. Recovery is an earnest leaning into the work it takes to move from trauma into behaviors that are additive rather than subtractive. It is taking common reactions to trauma and transforming them into intentional, positive responses like the kinds of physical, spiritual, emotional, and behavioral activities that rebuild and replace what's left.

My reluctance to recover comes from a lack of trust that my community and my job won't ask of me again what they've asked of me before. I struggle to let go of the past that keeps me tethered to readiness when every other billboard in my community says, "Be Ready, Butte!" with a flame in the corner.[4] To allow myself to recover from the repetitive hardship of fire, the aftermath, and my work in recovery, I must have faith I will not have to take this road again. And I do not have this faith.

Our resilience training addressed the trauma identity and the propensity for humans to hold on to who they become after tragedy. In the absence of their former complex identity, which trauma has replaced with a narrowly-focused state of survival, there may be nothing challenging their new belief that bad things will continue to happen. I know I don't want to stay in this damaged state of mind, but if I leave it behind and fire draws it back, will I have to climb another mountain as steep as the one I stand on top of now? I'm too weary for that.

---

[3] Pullen (2024).

[4] Be Ready, Butte campaign hosted by Butte Fire Department, CAL Fire, and Butte County. https:// bereadybutte.com/.

And yet, my persistent fear of recovering keeps me from living my best life. My lack of trust in a future rosier than one catastrophic disaster after another eats away at my soul. I find it hard to focus on working toward a vision of blue skies and green mountains when one plume of smoke on a dry, dusty ridgetop sends my nervous system into a frenzy for the rest of the day.

There are distinct differences even between rural, remote communities just a few miles apart if one has experienced fire recently and one has not. In communities rebuilding from fire, the conversation involves jokes about helicopters landing on rooftops so people can flee by air rather than worry about the roads. In communities recently threatened by fire but not burned, people talk about the one or two times they've evacuated in the past years and the ways in which they've refined their escape plan. In high-risk communities that do not have recent fire history, folks talk about the trailers they've parked in the valley so they have someplace to go when their homes burn. One woman living in an intact community told me she'd recently purchased a trailer to escape with her nine ducks to safety in the event of a fire, though she doubts the ducks will climb in when the time comes.

On Mother's Day in 2024, we gathered at my parents' house in the canyon and breathed in fresh air and enjoyed the quiet scenery of children climbing trees and hopping over rivulets draining from the cistern. I took in the fact that they're rebuilding their garage after five years, a mini house-like structure on their property with a pitched roof where my mom hopes her granddaughters will play in the rain. The very next day, I attended an evacuation planning meeting for my parents' neighborhood and heard an attendee call the area an "evacuation cul-de-sac." There's only one reliable evacuation route out of the canyon that, if blocked, will send people up gravel and singe-lane roads. Any obstruction may lock people—my parents—in the canyon with few to no options for escape.

I'm at war with myself and my own recovery, and with the peace I'm trying to make with my parents' decision to stay in the canyon. If I give in to recovery and the worst happens, will I be able to click back into this fragile but hardened state of readiness? If the worst happens and I have not recovered, have I done my best living in the meantime? If I don't recover and the worst never happens, have I squandered my life in fight or flight mode? The answer to every question is yes and no and, frankly, I'm getting tired of the question.

As much as I want to trust that the future does not hold another catastrophe, and as much as I want to trust in the sound construction of my parents' new garage to protect their precious belongings, I feel reluctant to do so. Our resilience instructors tell us to challenge our trauma identities with new experiences that lead to new beliefs and eventually the return of our complex selves. They gave us the tools to track our nervous systems, and to regulate and return them to a pleasant or at least neutral state after being triggered. They explained how the simple act of gesturing with our bodies can change a mood or interaction for the better, a strategy to adjust the inside by changing the outside. All of this makes sense to me, but is it enough if I'm barely getting by with daily tools and another fire happens?

Regardless if they're enough, they're all I have. I take breaks during evacuation meetings when the stories get too heartbreaking. In Forbestown, I stepped out into the mountain fresh air and took in the tops of the trees swaying in the spring breeze. I listened to the birds and let my imagination take me back to a time when nature wasn't so dangerous. During my regular work day, I try to take walks during each lunch break to reset my mind and connect to life beyond what's on my desk.

I do all the things I'm told to do, and read about, and listen to, to bring myself greater peace. I know I've come a long way from where I was a few years ago, and I've even built up some resilience. When the rage comes, it doesn't last as long as it used to.

I feel a measurable improvement in my attitude and ability to pause and rest in my skin after I flush strong reactions to negative experiences through my nervous system. I know I will survive the ride so long as I don't fight it. But I hit a snag when it comes to trusting in a different future, a better future where these beautiful, historic communities with swaying trees will never burn again. This is where I come up short.

I draw inspiration from kintsugi, which, as understand it, is the Japanese art of mending broken pottery with gold, rendering it more valuable than when it was intact. The fine art of mending with gold turns a shattered bowl, vase, or pitcher into a glittering treasure. I try to practice my interpretation of kintsugi with my mind, body, and heart. I am holding a bowl of gold and applying it to where I feel broken inside using a patient, unskilled hand. I know, however, that I won't feel more valuable to myself and to others until I trust the gold is strong enough to keep me from shattering again. On the flip side, I also believe that when I fall and shatter, there'll be more gold to make me whole.

My parents' garage feels like an act of mending their fire damage with gold, delicately stitching their property back together. Melissa's new house feels like a gold seam on her property for the way she built it for her future self as much as the woman she is today. For me to heal, to really recover, I need to shift into a sustained state of seeing the gold without seeing the cracks. I need to place my faith and trust in a safe future where Melissa will age in place in her lovely new home, my mom's grandchildren will play in her garage on rainy winter days in the canyon, and I will not crack under the weight of another disaster. Oh, how I want to believe that.

As I wrote this, an evacuation order popped in to my email.

While I hear myself wanting a different future, needing a different future, I am constantly reminded of the fragility of our current state and the fact that I'm very much living in it, inescapably, right now.

The Ranch Fire on May 15, 2024, caused the first evacuation order of the season but little damage. A series of small fires were put out rapidly by air retardant and ground response, and an individual was apprehended for suspected arson. At our evacuation planning meetings in the days following, we discussed the ever presence of these incidents and the need for readiness. A woman looked me right in the eye from the heart of the North Complex burn scar where she'd just rebuilt her house, and said if she lost another home to fire she'd leave Butte County and never come back (Fig. 10.1).

I heard the phrase "double consciousness" interpreted on a podcast to describe the simultaneous act of living one experience while creating another by posting about it

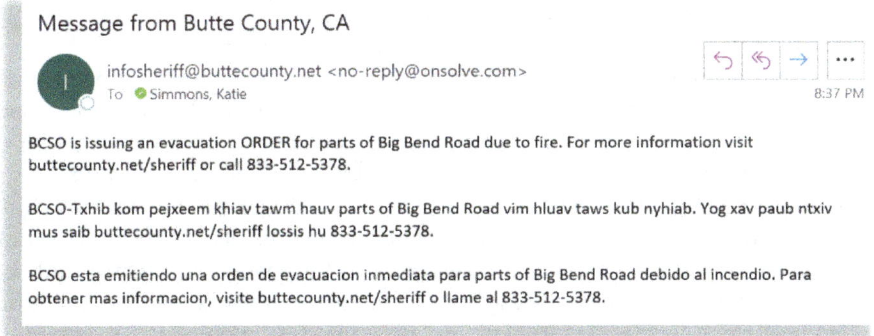

Fig. 10.1 Snippet of an email from the Butte County Sheriff's Office issuing an evacuation order

online.[5] Living in the real world and living online can lead to diverging experiences and circumstances, and the need to reconcile the space between.

In our resilience training we talked about trauma identity and the slow and sometimes regressive act of recovering and rebuilding. In the frame of double consciousness, I believe two trauma identities formed for me at nearly the same time, one at work and one online, creating two disparate selves to navigate.

During the pandemic, when I was working on recovery from the Camp Fire, just one month after the North Complex began, I was diagnosed with my genetic neuromuscular disease that I chose to share online a few months later. While I was fresh in the process of learning disaster recovery and experiencing ongoing fire, I was now learning about my disease and the future of my body, and spent a great deal of time processing these heavy topics online.

The Camp Fire created an "after" identify for me, one I would now consider a trauma identity once I embodied the Disaster Recovery Director title. Learning about my CMT also alienated me from myself. I felt betrayed by the body I confidently moved through the world in, that was now starting to fail. Choosing to engage in double consciousness which I didn't have words to describe at the time, had me running two parallel tracks in real life, and two very distinct narratives online. I am wounded; I am a warrior. My body is degenerating; I am unstoppable.

I justified sharing my clinical trial experiences online as helpful to others and clarifying for myself. Posting from the hospital parking garage after each appointment, before I had the benefit of reflecting on the drive home, meant I was sharing experiences I hadn't spent more than a few seconds feeling. And because I never cried with grief for this disease except for in that hospital parking garage on the way home, it was at the apex of my sadness and frustration that I manufactured my most hopeful outlook. Same with fanning the flames of my fire identity: posting hopeful

---

[5] The phrase "double consciousness" originated from William Edward Burghardt Du Bois to describe a sense of "twoness," due to racial oppression, Stanford Encyclopedia of Philosophy. https://plato.stanford.edu/entries/double-consciousness/.

sentiments from my devastating drives through Paradise in the morning as my morale bottomed out. In real life I was suffering; online I was cheering.

In resilience training, as we worked through the process of shifting our perception of our identities, I realized how many I was juggling at once. It is very clear and uncomplicated to live in the real world without feeling compelled to talk about it online. I am still very much a chronicler of my own life in writing, but I no longer choose to grapple with the reactions of others online before I grapple with my own.

We're learning that while a trauma identity is running its course, a series of interruptions, or turning points, can open the door for healing to begin. I told the instructor it's difficult for me to see a trauma identity until it's behind me. For him, moving past being a "cancer survivor" and back into his complex life made up of many roles—dad, coach, husband—was possible because of new experiences he chose to engage in that challenged his trauma identity. He fired his cancer doctor who limited him to his cancer identity by discouraging him from participating in a competitive sporting event after surgery. From my outside perspective, he's the one who got himself past the limitations of his trauma identity, and to hear him explain it, he sounds rightfully proud of the process.[6]

For me, as I've described, I am reluctant to let go of my fire identity because I feel like those wheels need to remain greased and ready for the next trek. With my CMT diagnosis identity, I am slowly moving past the limiting identification of myself as a diseased woman and focus on the things I can do. This is not a quick nor easy leap for me, but day after day I silence the inner voices telling me I am nothing more than the limitation of my disease. This doesn't mean I've jumped headlong into running again, because the physical limitations are real, but I do not curtail my expectations of myself to the bare minimum.

In real life, my online diagnosis identity compelled people to ask about my CMT. I'd regularly hear questions like, "How do you feel with your…what is it called?" Now, I breeze past their awkwardness and say I feel great, or head off those conversations before they start. Choosing to invite people onto my journey of a million miles before taking the first step left me managing their experience of my disease before I did. Now, I let people know quickly and kindly they're off the hook with this topic. It's no longer something we need to discuss or think about together. It's mine to carry as I continue to adjust to this degenerative disease, which is a lifelong piece of my identify, but not the only one.

There's a picture of me as a baby crawling around in the casts I wore for the first year and a half of my life to straighten my legs. Doctors changed the casts regularly to conform with my developmental stages. I had newborn casts, crawlers, then walking casts with tiny green blocks under my baby toes to create a flat surface for balancing. Most babies learn to walk on their feet, I learned to walk on blocks.

What strikes me about the photo is how happy I look, how unconcerned I was to be dragging the heavy, restrictive weight of plaster behind me. From my expression, I can see the casts weren't a burden at all, they were just attachments I took with me across grandma's green linoleum floor. This photo reminds me that though the

---

[6] Fellowship interview, Former Instructor for Butte College—The Training Place, 2024.

burdens we carry are very real, they don't have to keep us from moving forward with smiles on our faces. My goal now is to relearn what I knew as a baby: this disease is something I am strong enough to carry with me wherever I go.

The collective trauma identity formed in our region through experiencing multiple disasters is likely here to stay for a while. Our evacuation planning consultant asked about the Oroville Spillway Incident recently, wanting to understand why the dam was rebuilt rather than torn down. In conversations like these, I hover above the scene and watch from a distance. I watched my colleague relive the memory of that incident, and the impact of the nation's largest mass evacuation on the community and County staff. As we stood there in the North Complex burn scar working with the community on affirming routes and assembly points, we were transported by retelling the story of yet another recent disaster.

The sliding doors between the Spillway Incident, the Camp Fire, the North Complex Fire, and the Park Fire, are barely visible except to those who've lived through the unique experience of each, and carry the collective trauma of it all. It's too fresh and significant for us to shrug away even one question because of how desperately we wish to be seen and understood. We pay the emotional price to hand out our valuable insight so that others may explain on our behalf in the future.

As well-meaning as casual questions about our disasters are, we address them constantly which keeps them on regular rotation in our minds. And while we handle these interpersonal exchanges well on the outside, the toll they take on us is largely unseen because it's impossible to measure the frequency of re-traumatization after disaster.

The good news is, with training and time, resilient choices feel more and more possible. Knowing I can challenge my fire identity by choosing experiences that expand and enhance my outlook in different ways puts the wheel back into my hands. I appreciate that I know wildfire and recovery, but I want to know more, I want to experience more, eventually I want to write about much more than this.

10.2 Animal Healing

At a department staff meeting just before the summer of 2024, while we talked about how our personal values align with our mission, a co-worker told the story of being one of the first staff members to cross the fire line during the Camp Fire. He was dispatched early to deliver food and supplies and was allowed to cross even before the National Guard arrived. He described the large animals he saw as soon as he entered as, "newly scarred and still on fire." The fact that he mentioned the animals before talking about the people who were grateful to see him made me wonder if he has big animal pain, too.

Two years ago, my youngest daughter, Heidi, started begging for a dog. We took her to the library to check out dog books to research the many lists she was making at home. As she studied her library books, she wrote down dog breeds in two categories:

breeds she wanted and breeds she did not. On the list of dogs she wanted, she included the largest possible breeds which I nixed.

To play along, I half-heartedly contributed a few ideas to the list but was fully against the idea of getting a dog. We have two cats, numerous fish tanks, house plants, and plenty of other responsibilities. I was not in favor of adding a dog to the mix, especially not a big one. After two years of listening to this, however, my husband and daughter broke me down and I told them the decision was theirs.

A while after the dog lists started, our oldest moved out of the house for college. Heidi was alone except for our cats and we could tell she missed being playful and silly with her sister, as much as we tried to entertain her. Something rare and sacred was missing from her life.

We researched volunteering at the local humane society to satiate Heidi's dog obsession, but the shelter requires kids to be a minimum of twelve years old and accompanied by a parent. At that point, Heidi was still months away from turning from 12 but was eager to visit the shelter and desperate to get a dog. When we agreed to let her look, she was thrilled.

On a Saturday afternoon in April of 2024, we drove to the shelter to look at the dogs up for adoption. The idea was we'd visit the shelter over the coming months and years to see if she connected with a dog that might be suitable for our lives.

The noise was jarring when we stepped outside to the kennels where dogs barked at us from both sides. I was ready to leave right away. A staff member already helping another couple let us tag along for the tour. We walked down the central path with kennels filled with dogs jumping and barking, some wearing cones. We turned left and viewed another set of kennels before moving to the furthest row on the right. There, in the very last kennel of the very last row was Sadie, tall, white with brown ears, totally silent and unmoving.

While the rest of the dogs jumped wildly in their kennels, Sadie came to the fence to greet us, but as we moved closer she turned and went inside. We noticed how quiet she was and asked a few questions. The staff member told us she was about four years old, a Great Pyrenees mix, very mature for her age, quiet, and loved tummy rubs. We waited for a minute and out Sadie came again, as curious about us as we were about her. Heidi was beside herself. We told the staff member we'd like to meet Sadie.

Since Sadie had just been spayed, we had to wait in a room inside the shelter to meet her rather than out in the yard. We sat and waited until she swished in on her long legs, a huge bundle of curious affection, reaching out with her paws and nose to touch us. At one point, she easily placed both of her paws on Rodney's shoulders and looked right into his eyes. Heidi said, "I'm in love," and my husband wasn't far behind.

I wasn't too sure about Sadie but I know when I'm outnumbered so I began to prepare myself for the inevitable. We met one other dog, Penny, who barely noticed us for the distraction of the other dogs outside the pen, and we left the shelter talking about Sadie.

The lobbying started as soon as we got to the car. Heidi and Rodney were really excited and spent the rest of the afternoon pricing out all the things we'd need to buy if we adopted Sadie. With the adoption fee and large dog supplies which are 3 times the cost of small dog supplies, the shopping list put us well over budget for the month. But Heidi was in love so we researched Great Pyrenees dogs, filled out the application, and told the shelter we were thinking about adopting Sadie.

The next day, Heidi and Rodney went back to visit Sadie who was still a sweetheart, then called me from the parking lot. Serendipitously, the moment we decided to adopt her, the shelter called and said our application had been approved. They picked me up and we went back to the shelter and adopted Sadie. She walked straight out the door and climbed into our car as if she'd done it a thousand times.

The first few weeks adjusting to a dog weren't easy. Sadie was instantly comfortable in our house but barked occasionally "with authority" as our neighbor noted, and had extremely poor breath and body odor from living in the shelter. We couldn't bathe her due to her recent surgery so we kept our doors and windows open.

It's fair to say Heidi was over the moon from the moment we adopted Sadie, while my husband and I vacillated between overwhelmed and reluctant. We'd settled into a nice rhythm during the school year and having a dog interrupted that flow. Our first full day with Sadie was Monday and while I went to work and Heidi went to school, Rodney set about trying to figure out a routine for a 70-pound rescue dog who'd just had surgery. She was a little too interested in our cats for comfort so we had to get creative with space for safety. We blocked off our bedroom with a wooden kitchen chair so the cats had their own space, which meant we had bruised knees for weeks.

Over the next several months, with some dog training for Sadie and for us, she settled in and we found a bit of a groove. When we first took her home, we launched into three-mile walks twice a day when, as it turns out, she prefers walking one mile twice a day at most. We'd leash her up and bring her to community events in a semi-panic not knowing how she'd respond, when now we know to crate her while we go out so she has the downtime she needs and enjoys.

We've learned Sadie likes soft bacon-flavored bones but not aggressive chew bones, and she'll do anything for freeze-dried chicken treats. Our cats aren't totally on board but my husband can't imagine life without her and I'm becoming attuned to the rhythms of having a dog. Sadie is happy as a clam who, yes, wants her belly rubbed constantly.

As we worked through the tensions and challenges of adapting to life with a giant dog, I noticed something changing within me. On my way to work in the morning I was able to drive by the grazing cattle without that familiar tug of pain. In fact, there was an absence of pain and space for something else: peace.

Falling in love with a big dog who looks like a horse-cow, who can stand on her hind legs and reach around my neck to give me a hug, settled something in my soul. A few months in to dog parenthood, I told Rodney I believe Sadie is healing my big animal pain. As cattle trucks roll past on the highway transporting cattle into the mountains for the summer I am not filled with fear and grief. Instead, I take those moments to feel peace and hope for their safe passage and an uneventful fire season.

Sadie has gained 45 pounds since she arrived, and fills the entire kitchen floor when she lays down to rest. She's too big to pass in the hallway so I have to walk behind her no matter how slowly she goes. I watch the brown spot on her back sway back and forth as she lumbers along, and wait patiently for her to get to where she's going so I can get to where I'm going. She's not a high energy dog and does everything on her own sweet time, sometimes collapsing in slow motion from hip to ears to expose her belly for pets. The moment she starts barking I hold up her brush and she's on the ground, waiting.

My animal healing is coming from the growing love I feel for Sadie and the love I feel so easily coming from her. She trusted us instantly and hasn't looked back. She lays in the whining chair every morning when the sun rises to watch cartoons. She is leash trained, loves her post-potty-break treats, and plays chase in the backyard with Rodney and Heidi, outsmarting them with her galloping pivots and unexpected full-stops. When we're not looking, she digs big holes in the lawn and eats bird food. She's too big and strong for me to hold on a leash but I love to go for walks with her and Rodney and Heidi, our family a different kind of foursome.

I hope the other animals in the shelter find their families, too. I hope the high energy dogs find high energy households with lots of space. The moment I get too sad for the dogs in their kennels I think about the one we were able to bring home and give a peaceful life. The simple act of sharing our lives with one dog who needed a home feels like healing for all of us.

Sadie has taught me that recovering from deep pain is possible, because I sense she's doing it, too. Feeling my big animal pain subside gives me hope that my tree grief will also ebb over time, maybe through another unexpected miracle as significant as a rescue dog. With Sadie in our home, I am not who I was before disaster recovery work, but I've moved several steps closer to a new normal. Adapting to her and our dog-family routine is an experience that is challenging my trauma identity. I can tell the more I love Sadie the more space she frees up for peace.

Reference

Pullen S (April 22, 2024) Butte County DA's office honors national crime victims' rights week. Chico ER. https://www.actionnewsnow.com/news/butte-county-das-office-honors-national-crime-victims-rights-week/article_7b905d8a-00f7-11ef-8a64-835f32facd15.html

Chapter 11
Surviving

It's possible I have arrived at the most painful part of this book, the part where the story simply isn't true if I don't share the rest. Like Dave Daley's cow story, I haven't found many safe places to share my own pain, so it persists.

At the very end of 2021, I cut my long hair into a short pixie. It was supposed to be an act of liberation as many women find a bold haircut freeing, but it wasn't for me. It was the start of the end of something which was almost too painful to bear. With my short hair, I felt the last of my old self, pre-disaster and pre-diagnosis, disappear and there was no one to catch me on the other side. My social anxiety grew to the point that I fled social gatherings and meetings, texting apologies as I went, unable to describe even to myself what was happening. All I knew was that I wasn't myself. In fact, I felt like no one at all.

With my short hair I went hiking and posted selfies of my adventures, staring into the photos after they were online, searching for a spark of recognition. It was the moment my double consciousness bisected my life completely. Online, I was free, liberated, climbing mountains, overcoming CMT, thriving in spite of it, my face free of make-up, looking directly into the lens. In real life, I held my phone and searched my own eyes staring out of Facebook for some sense of self, reading the comments as evidence I was visible to others. If I had proof everyone else could see me, then maybe I was there.

I got a lot of compliments on that haircut when I was brave enough to go out. In some strange burst of false confidence, I'd say it was the style that made me feel closest to myself. At night and in the morning, I'd sit in front of my closet and instead of seeing my clothes I'd see Katie's clothes. There was another woman living in the real world who wasn't me either, though my girls would ground me temporarily by saying, "mom" when they looked at me.

I've thought a lot about my extreme disassociation from my body, my self-image, and the person I portrayed online. Was it a result of the fires, the diagnosis, the pandemic, family challenges, the switch from my comfort zone of non-profits to the complicated universe of the public sector, all of the above?

K. Simmons, *Three Fire Mountains*, https://doi.org/10.1007/978-3-032-17343-0_11

During this time, I was on medication for CMT because I said yes to everything my doctors suggested after my diagnosis. I agreed to take the only medicine currently prescribed for chronic nerve pain which I consider the least of my concerns today. As the divide between my selves grew farther and farther apart, with no longer even a thread connecting the two, I said no to those pills. In fact, I was scrolling through a CMT support group on Facebook late one night when I saw how many people had experienced suicidal ideation while taking the same medication. This was all the validation I needed.

The next morning, I started to taper the dose until I finally stopped taking it all together. It's possible the medication was a big part of the problem because these are known side effects, but because I was halfway through the clinical trial at that point, I'll never know the extent of it. All I know for sure is that ending the medication was not the simple fix I'd hoped for, but it started me down a necessary path.

I've come to see that time of my life with a little bit of hindsight and, after characteristically drilling into the question of why, the reason doesn't matter as much as what I choose to do with it. The medication taught me that if my mind is sound, it matters less what's happening with my body. That's an invaluable lesson for facing a degenerative disease that I'm sure will play into my choices from here forward.

Hitting bottom could've been caused by any number of things: the fires, joining the disaster recovery field, the alienation from my body, the isolation from my community during the job changes and pandemic, the friendships that didn't go the distance. I've spent a lot of time looking for the one thing I could fix to make the hurt go away, but I don't think that's possible anymore. A resilient mindset takes all available building blocks and makes something new.

11.1 Defiant Resilience

In my twenties, I experienced my eating disorder as grieving the fragility of life while still alive. Recovering from the disorder felt like the gradual waning of grief while my life was restored. Recovering from disassociation and depression feels similar. As my two closest friendships were winding down during the worst of it, I tried to be positive and grateful I could articulate some of this to them so that I might explain. But it was soon clear I could not ask anyone to pause their lives for me. I had left my life and there was no catching up on theirs.

My clinical trial tested for suicidal ideation and I gave the answers that felt true to the patient who wanted to succeed as a research subject. I have no shame in that because of how desperately I needed to be part of something meaningful. But my desire to succeed probably kept me from getting the help I needed which is not an uncommon theme for me.

I tried a few times to form new friendships but I had nothing to give, and I'm far too prideful to be a taker, so I let those people go, too. At work, I withstood criticism from colleagues who didn't think I was strong enough and smart enough for disaster recovery, and though they might have been right, my leadership believed in me.

Looking back on this all of this now, I see myself on the side of an icy cliff, pick-axing my way up one swing at a time, sometimes missing, slipping, then holding on for dear life before anchoring again. It sounds dramatic but these are the only words I can find to describe the white-knuckle experience of thinking maybe I shouldn't be alive anymore because I was already gone.

I credit my husband for being solid even when our relationship wasn't. I credit my daughters for being unapologetically themselves and showing me by example that space and time and resources are meant to be used for strength, healing, and growth. During the worst of the disassociation, I'd be surprised when they'd flutter around me at night telling me about their days, when I couldn't even feel myself in the room. I'd see them seeing me and use that to be present. Thankfully, my mind is back online and I can stay inside of my body easily enough to hold any conversation. I don't always feel great, but I'm very much in the room.

When I started the fellowship, I didn't plan to talk about suicidal thoughts, or my dad's death, or CMT. I planned to talk about fire through the narrow lens of data, using it to show how policy can be adapted to help set the pace and scale of recovery. As it turns out, reckoning with fire is indistinguishable from reckoning with life, because fire touches every part of life, and my life will never be the same. Turns out I am the research subject for fire as well.

I am proud I pursued the fellowship on wildfire recovery while still engaged in it professionally. I stared into the barrel of slow recovery and decided to unpack that challenge so others may learn from our losses and lessons. I hope shining a light on my personal journey with the fires and disaster recovery will help complex people carrying a variety of roles—spouse, parent, friend—feel less alone if wildfire and recovery derail them, too. Disaster recovery doesn't occur in a sealed test tube controlling for all other variables, it explodes violently out of the laboratory. Life goes on, messy as it ever was, probably more so.

I miss the person I was when I didn't worry about my house burning down. I miss looking at trees and seeing great bastions of nature instead of vulnerable fuels for fire. I miss the innocence of watching cows graze, and loving wildlife instead of grieving animals desperately after fire. I miss the professional I was when I flitted bravely in front of people, leading meetings and events with the assuredness of a woman who believed in herself. I look at her now and wonder where she got that swagger. It doesn't feel authentic anymore and it's a little bit embarrassing, but I hope the cracked version gleams with gold.

I closed the divide between my online presence and the real world by deactivating my social media accounts. No more staring at myself trying to get reacquainted, and reading comments to feel alive. Removing myself from the internet was necessary to regain an awareness of my physical presence in the world.

The good news is that though I'm fully embodied but not totally at peace, I don't feel shame about that and rarely judge myself anymore. Instead, I try to show up with my real voice and give myself permission for it to shake. If nothing else, this chapter has taught me I can survive not being liked or belonging, as long as I have a true and solid self to fall back on. People pleasing is far less compelling than it used to be.

The process of writing everything down is also an act of finding my voice again. It is the ultimate interference in the isolation of disaster recovery, maybe even an act of defiant resilience. Through writing, I can feel myself disconnecting from some of these experiences enough that I never want to describe them again. What a blessing it is to let go.

11.2 Closing the Gap

Over the past few years my hair has grown back. I can now pull it around and tuck it in front of my shoulder like I've done since grade school. Driving down from meetings in the foothills, my hair blows out the window and makes me feel lighter and free, a little piece of the old me coming back. As trite as it sounds, looking like myself makes me feel more like myself.

As I write this, I realize what I've been searching for at work is the same thing I've been searching for inside. The definition of recovery is important at work because it gives us bearings in space and time, it is also important to my ability to gauge how I'm really doing. While I've used this research to explore the many definitions of recovery on paper, I believe I've found the essence of it inside of myself which brings me back to the obvious.

Recovery isn't static and defined, it is constantly changing and moving and perhaps teasing just out of reach, a reason itself to keep reaching. It is the act of putting effort into something, rather than the effort itself. Perhaps the federal government is right when it describes recovery as the capacity to keep going rather than a state of arrival, because it isn't a fixed object we can measure our distance from so we can calibrate how much further we need to go. It is the messiest, ugliest, exhausting middle you can imagine that never seems to end, the hardest stuff you can dream up, the most enraging. It is the art of repairing what's broken with gold.

Recovery starts with animals on fire, eventually indistinguishable from ash themselves, and heartbroken ranchers who start again. It is trees engulfed in flames, gradually cooling from the heat into new shapes and sizes, to stand as burn-scarred sentinels in our black and white communities, crumbling slowly or falling swiftly from a gust of wind or an axe, or allowed to stand with a whisp of canopy into the next season to seed a new generation. It is each one of those animals and each one of those trees, not the sum but the individual parts that add up to the infinite value of being alive.

Recovery is homes bursting with fire then smoldering in wreckage, scraped and sprayed during removal by tank trucks with water hoses, until nothing but a caged pool bottom is left, a place for a family to start over or start anew, or a place for wildflowers and brush to grow while the land turns back into itself.

Recovery is people shaving in parking lots then being rushed to the hospital, working, panicking, adrenalized beyond recognition, listing their lost belongings in an offensive act of putting a dollar value on irreplaceability, then pausing for long months or years in the best housing they can find, starting down one path before reversing into another, bewildered for however long it takes to find home, or forced to define it differently for the rest of their lives.

Recovery is the workforce with destructive images burned as deeply into their minds as the fire itself, showing intangible things like peace and trust are just as flammable as the objects outside, those workers finding people and places with which to heal, or ducking into a long and tenuous grief that closes off their complex selves beyond the one identity that remains, until the sun rises and sets enough that the pain becomes a thing they carry rather than the other way around. So many people lost their lives in the aftermath who never got the chance to heal. If we're keen on what fire takes, we can see it takes much more than it burns.

Still, like the trees standing tall with smoky black trunks, there are giants in our community who bear their scars and lean even further into the hardship because they have small miracles to offer the heartbroken. Maybe those miracles are memories they can share, or ideas for new policies, or muscle and time to help someone rebuild. There are many people in our communities who are using their bodies and their brains to swing hammers both figuratively and literally at the beast in our neighborhoods, showing it they're much stronger and wiser, or at least unafraid. Or they're like me, invited in without much to offer and finding just enough along the way to keep going. This is recovery, too.

I wish I could put a period on the end of this book and catastrophic wildfires would never happen again. But as the headlines say time and time again, there are wildfires burning out of control somewhere, and our first responders are gearing up for a fire season no one sees the end to. In Butte County, we have an updated set of evacuation routes and assembly points to use in case of an emergency, while foothill residents are parking their RVs in the valley as Plan B.

I asked Jovanni Tricerri who worked for our local foundation in the aftermath of the Camp Fire to define recovery for the fellowship. Having worked with survivors and agencies directly supporting recovery, he wrote:

> "I often said that while disasters do not discriminate, the recovery process often does. Recovery, to me, transcends mere physical rebuilding; it's about ensuring an inclusive, equitable process that actively supports the most vulnerable. The goal is not only to repair what was lost but to enhance community resilience and cohesion, ensuring our more vulnerable neighbors are given extra efforts to access resources and opportunities."[1]

Vulnerability can be plainly visible like it is in our recovering communities, or it can be protected from view as it is in many of our survivors who are just getting by even if they look ok from the outside. When we talk about extra efforts in recovery, we're referring to more than securing housing, employment, education, food, services, sustainable financial stability, the bedrock of meeting basic human needs. We're talking about belonging, acceptance, celebration, joy, hope, pointing our goals toward achieving these things so that we may all thrive.

---

[1] Fellowship interview, Jovanni Tricerri, formerly with the North Valley Community Foundation—Butte Strong Fund, 2024.

If down payments are made after disaster to test community viability, and we stop there, then we are settling for survival at best. Viability by its very definition is the ability to live, which is important for the deployment of recovery resources from investors like the federal government, but not enough to restore quality of life.

11.3 The Future

Our resilience instructors explained the difference between solving problems and following a vision. Solving a problem requires focusing on the problem, while following a vision does not. Recovering from disaster is different than envisioning a new community, and healing from trauma is different than living our best lives. On one hand we're restoring what we've lost, on another we're building toward a new future as new people. In a perfect world, we get to do both.

Repairing breakage with gold represents an art that can't exist without damage. I am acutely aware that I am trying to solve the problem of the Camp Fire and heal myself from trauma, rather than creating a vision for the future. The art of repairing my wounds by writing about them frees me up to focus on a future not solely defined by damage. Much as I saw suffering in the eyes on that magazine cover decades ago, I see suffering in my writing. The story is outside of me now and I can read it, learn from it, and hold it far enough away that my pain becomes separate from my body.

Suffering is within all of us in this region as we navigate from the worst that could happen into a future that can only be reimagined with a bit of fearless vision. Just as Heidi said, I don't need to carry the fire forever, so with this writing I give it wings.

Heidi and I went on a bike ride up into the foothills above her middle school recently. On a smooth stretch of downhill I told her to watch me as I lifted my feet off the pedals and kicked my legs into the air.

As I did it, I realized pieces of the old me—the little "speed demon" my dad watched flying down grandma's driveway, and the young woman in Paradise finding her freedom on a bicycle—has made its way into the present. I told Heidi to do the same and pretty soon she went flying by me, giggling, kicking up her heels.

11.4 Reflection

Every fire is different. Every recovery is different. My hope is there are a handful of useful observations in here for anyone dealing with fire, aftermath, and recovery, personally and/or professionally. This book is not much more than a memoir, and it's not meant to be instructive. It is an honest offering of experiences and opinions that may dispel some of the complexity and confusion of coping with fire.

Clearly, I am not a recovery expert. I have not been formally educated on response and recovery beyond what I have learned and interpreted on the job through my own lens, and the trainings my employers have generously offered for our professional development. I encourage formal training and reading recovery manuals, and I strongly believe anyone working in and around recovery should question and probe at historical practices so the field continues to evolve and improve.

I do not share my story lightly, and I fully acknowledge that many people in my community and other communities impacted by wildfire have far more harrowing tales. I grieve the lives lost to our wildfires and hope that though I have not honored them individually by name, I have pushed forward these stories so that others may learn.

There's no end to this book because there is no end to fire. The recently retired Fire Chief for the Moraga-Orinda Fire District, said on a call recently, "fire is a feature, not a bug." Fire is on our landscapes and in our lives. It is now a permanent part of who I am. The best we can do is listen, learn, iterate, challenge, and keep trying to stay one step ahead.

Afterword

I finished my fellowship research on June 21, 2024.

Apache Fire, June 24–29, 2024

Thompson Fire, July 2–8, 2024

Grubbs Fire, July 3–4, 2024

Railbridge Fire, July 11–13, 2024

Park Fire, July 24, 2024–September 27, 2024

On the day the Park Fire started I was monitoring my phone because that's what we do now. We watch and wait for the tiniest twitch to tell us it's time to report to the Emergency Operations Center. I saw little more than where the fire started and how fast it was moving before texting my husband to pick up our then 11-year-old camper from the park. It's hard to say if it was mother's instinct or a readiness I can only describe as "the world is about to burn down."

He did pick her up and, as it turns out, that was her last day of summer camp in 2024 and the first day of the 4th largest fire in California history to date.[1]

It is important to mark dates because fire timestamps life; it also erases time. Fire locks communities in a state of suspense while doing its thing—for days, weeks, months—then hands everything back and shrugs like it was never there. And except for the destruction it has wrought and the landscape it has altered, the sky turns blue again and life goes on. For the people working through it, or living through it, they've been catapulted into the future through the sliding door of fire.

After fire, life devolves into the clunky mechanics of every burden imaginable for survivors. The smoothness of a life undisturbed by fire is something we take for granted until it is gone. As friendly as the staff are, the last thing anyone wants to

---

[1] Reported to me by numerous first responders when the Park Fire was contained, 2024.

K. Simmons, *Three Fire Mountains*, https://doi.org/10.1007/978-3-032-17343-0

do is push their walker through a Local Assistance Center gathering fliers for social services after losing everything they own.

Before the Park Fire was contained, I heard a consumer advocate express relief that it wasn't as destructive as the Camp Fire. I emailed her that a fire that destroys a home is the worst fire that homeowner has ever experienced. There may be inequity in the scale of fire, but there is equity in destruction for the people impacted. The following day, in our evacuation planning meeting, a public safety official commented that by day two, the Park Fire exceeded the total acreage of the Camp Fire which burned for several weeks. If we're measuring destruction of habitat loss by volume, the Park Fire exceeded the Camp Fire by a factor of three. The Camp Fire, however, took far more lives and structures.

So, let's not compare fires. They all hurt in different ways and in different places. They hurt our homes and our brains, scarring our memories. We lose places and, tragically, people we will never get back from death or from the walking death of recovery.

The Thompson Fire started on July 2, 2024, right outside of our office building. As it picked up steam, billowing and blowing wildly, we were called into the Emergency Operations Center. I carpooled with a co-worker, steeling myself for the fear ahead, and the responsibility of reporting to work for an indeterminate shift on fire.

On our way to the EOC we passed kids on the street looking up at the smoke curving easily over their heads from the hillside, swallowing their summer sunshine and shading mid-afternoon. Fire is terrifying in the distance and incomprehensible up close. As we drove closer to the flames, I prayed those kids had an adult nearby to whisk them to safety. Even as I did that, I knew fire was imprinting itself on their memories as something they'd never unsee.

From the EOC, we watched the fire wind whip the trees and smoke layer and color itself in light and dark. I took a video for my family so they could see from their relative safety the utter chaos we were experiencing. Within hours we relocated the EOC from Oroville to Chico, where we, too, could be safe from this one.

The 4th of July came and went during the Thompson Fire, one day, the next day, no days the same but not unique enough to remember, because time has a way of stitching days together during fire.

At its peak, the Thompson Fire displaced well over 10,000 people temporarily, ultimately destroying 26 residences and structures, and damaging another eight.[2]

Throughout the fall and winter following the Thompson Fire, the view of the Oroville hillside from my office was ringed with black char. The big white "O" for Oroville popped out of the blackened cliff like a stunned eye. Driving into work after the Thompson Fire the air was resonant with smoke and ash—not the dank distant smell of fire but the close, sharp presence of 3,789 acres of ash and soot. We blasted our air purifiers until our noses blinded themselves to the smell.

The Thompson Fire was pockmarked by other fire starts, four in one day. The noise of the EOC hummed along unbroken as our phones beeped and buzzed with

---

[2] CAL FIRE Incident Report, Thompson Fire, last updated July 8, 2024, https://www.fire.ca.gov/incidents/2024/7/2/thompson-fire.

news of other fires. We added new evacuation zones to the whiteboard, opened a new tab in our minds for a second set of fire stats, and moved on to the next task. The Grubbs Fire came and went, taking 10 acres and at least 1 structure. We held our breaths the night of the 4th, hoping errant fireworks wouldn't rain on this terrible parade.

CAL FIRE Incident Management Team 6 fired up the Fairgrounds in Chico for the summer, moving in their trailers, tents, and trucks for the Thompson Fire. While personnel were out on the line, we took a quiet afternoon to tour the Incident Command Post as a training opportunity for Butte County staff new to emergency response. The trailers were air conditioned because temperatures skyrocketed well past 110, bearing down from the sun and radiating back up from the asphalt in boiling waves, cooking the smoky air. I was standing in the 0700 briefing one morning when the operations chief warned of temperatures reaching 118 and I excused myself to sit down, wobbly at the thought of everyone around me surviving the heat and flames. "This is stupid," I said to myself over and over, disbelieving their circumstances.

Another Big One

A few weeks after the Thompson Fire tapered, we got news of a fire start in Upper Park in Chico. I took one look at the dirty brown column and texted my husband. I worried I might be overreacting but then he and the rest of the community watched as the Park Fire exploded in the foothills, hell bent on becoming another record breaker. The Incident Commander clocked the fire at 5,000 acres per hour without wind during those first several days.

We reported to the EOC in Oroville just after our workday ended on July 24, 2024, and began a shift that turned into disaster recovery that lasts to this day. The Park Fire is the 2nd largest single-ignition fire, 4th largest overall, in state history until another fire brazenly stakes this claim to fame. The two largest single-ignition fires in California started in Butte: the Dixie Fire at nearly a million acres and the Park Fire.

Incident Management Team (IMT) 3 was assigned to the Park Fire initially, then refocused on the Butte Zone as the Tehama Zone stood up and IMT 4 arrived. Our IMT Liaison jokingly referred to the unified command of Team 3 and Team 4 as "Team 7." 400,000+ acres is a big fire requiring two Incident Command Posts: the Fairgrounds in Chico and the Fairgrounds in Red Bluff 40 miles away. Chico remained headquarters until the fire took itself to Tehama and settled in for the remainder.

The Park Fire was plume-driven, meaning big enough to create its own weather system which ultimately self-implodes. Above the smoke cloud, pure white cumulous clouds form—ice caps, they're called—indicating fire weather. The Incident Commander explained, "Think of a plume like a watermelon falling from the sky, fire splats all over when it collapses."[3] Plumes are what fire looks like when it takes a big deep breath of sky, gathering energy and steam for its next move.

The sight of fire holds whole counties hostage with its otherworldly power, showing people there are phenomena far stronger than their bodies, equipment, aircraft, and incident management systems. When fire is out of control it flagrantly brandishes all of its cards, hiding nothing as it wreaks its crazy havoc.

One night, out on a walk between EOC shifts, I had a conversation with the Park Fire. I could see how upset she was, how vengeful. In my small body with my small brain I told her we could see her, feel her, acknowledge her anger. She was too busy to speak back to me so she carried on, leveling her plume before nightfall and glowing red from the ground. I attribute my Park Fire talks to the zombie-like sleeplessness of response, but I also believe communities go to a place far beyond reality to comprehend fire. We barter, we beg, we try to reason with the beast. As powerless as we are, survival instinct has us whipping every tool out of the toolbox. I sense this same determination at the Incident Command Post day in and day out: steady, informed, reasonable, repetitive, predictable, aware, constant.

The Park Fire nearly brought me to my knees one night as I sat behind my desk at the EOC. Evacuation warnings were called in Magalia and Paradise, communities still recovering from the Camp Fire. The thought of the Park Fire doubling back to take communities already taken by fire seemed wildly inconceivable. As we coordinated our County response to this threat, I felt my own body threaten defeat. I heard my voice strong and mechanical, while I knew that if this fire burned Magalia and Paradise, I would need a long, hard rest on the floor. Face down. For months.

That night, as I drove home to stay alert to my phone in the half-sleep of response, I saw the Park plume glowing pink and red like the open mouth of a volcano. It sat unmoving above my town, above our foothills as empty of people as they could be, as full of wild animals as they'd ever been, and I said, "enough, fire." You. Have. Taken. Enough.

The Park Fire decided not to double back, but it didn't stop doubling down. It found the Mill Creek Drainage which IMT 3 calls, "legendary in the firefighting world." The drainage isn't a small, dry creek bed as its name would suggest. It is several miles wide, the perfect funnel for fire to move unimpeded, unreachable, unseen when its plumes ground aircraft. By the time the Park Fire established in the drainage it had left its mark in Butte County, destroying 408 structures in Cohasset, 62 in Forest Ranch, and 5 in Richardson Springs.[4] The familiar tide of slowly learning of friends' losses ebbed and flowed. I surprised myself by bursting into tears at the sight of a friend who lost her beautiful home in Forest Ranch. Weeks of exhaustion and grief poured out in that single hug.

Our Emergency Operations Center began deactivating as we activated the Disaster Recovery Operations Center (DROC), which is the structure that stands up early recovery processes like debris and tree removal. For a while both the EOC and

---

[3] Heard directly from the Incident Commander on the Park Fire, 2024.

[4] CAL Fire Incident Report and Damage Inspection Report, Park Fire, last updated July 29, 2025, https://www.fire.ca.gov/incidents/2024/7/24/park-fire.

DROC were operating concurrently, on top of regular life and work, which is the co-occurring nature of late response and early recovery. I tried to keep pulse by letting cooperators know we were focusing on response, repopulation, and recovery to demonstrate the layered work county government does when fire moves through.

When I was ready to tour the Park Fire damage for the first time, our Fire Chief drove me and a co-worker into Cohasset, narrating the scene and inviting us to walk on crispy overlooks to take in the magnitude. Ahead of us in another truck were two other colleagues and the Battalion Chief who briefed the EOC daily on both the Thompson and Park Fires. We stopped in various places to see what survived and what didn't. At one point, staring out into the black, white, and gray world, I asked the Chief if fire blindness is a thing and he said yes.

Above the site of a former antique store where hundreds of classic car hulls sat rusted and ashy, the trees stood frozen by fire wind, their pine needles still faintly green but crisped and pointing away from the canyon where the fire bore down. Before they fall or are removed, trees are the weather vanes of fire.

My eyes are accustomed to spotting these details in a burn scar now, trained to know the direction and heat of a fire from what's left behind. We got out of the truck at the canyon edge where nothing but white ash piles and black tree stalks stood for several thousand acres. A peep peeping turned our heads to a covey of quail hopping quickly by. Life!

The roofs of standing homes were coated pink with retardant, green shrubs and bushes still growing under the windows as if nothing had happened, contrasting piles of metal, a skeletal chimney, and wire whisk trees next door. Fire shrinks homes to nearly nothing, making it impossible to know the scale and size from the ash pile left behind.

The Chief pointed out green shoots pushing up from the black claws of charred manzanita and oak brush. Oak trees are hardy in fire he explained, as we've learned, and will surprise us by coming back to life a year or two after fire. But that won't be the case for pine forests where animal carcasses were removed before we arrived, and where, as we experienced during the first rain following the Park Fire, soot turned to sludge and sheeted down the canyon walls into waterways.

Predictive Questions

As we build our data to support Park Fire recovery a few key questions stand out. These questions are in the Triage Tree and I believe are some of the most critical to understanding the complexity and length of recovery. Complexity draws out recovery which prolongs suffering. To me, recovery isn't best approached by answers—because trusting the source can be dubious—but by knowing what questions to ask:

Length of recovery:

Short   *Long*

|———|

How many fire survivors were renters?

None	Some	Half	Most	All

How many fire survivors were on fixed incomes?

None	Some	Half	Most	All

How many fire survivors had inadequate insurance?

None	Some	Half	Most	All

How many fire survivors lack access to municipal water and wastewater?

None	Some	Half	Most	All

What is the general housing availability for fire survivors in the region?

High	Moderately High	Moderate	Moderately Low	Low

Is homeowner's insurance available and affordable for reconstruction?

Yes	Likely	Moderately	Unlikely	Not at all

Questions will vary by disaster and by jurisdiction, but developing processes like intake surveys to gather survivor information at Local Assistance Centers, and tracking key indicators can generate critical information early in recovery. Understanding the length of recovery will help government staff and elected officials estimate end dates on key policies that govern recovery activities like urgency ordinances, and calculate the resources needed to fill any gaps.

Impossible Love

I had the opportunity with this research to develop a predictive model of recovery to determine whether or not a community might achieve viability after wildfire. I got pretty close to making this my sole focus, convincing myself that viability is the goal because that's what the scholars say. I told myself to look from the inside out, focusing on what (or who) is most likely to recover, then calculate if that's enough for the majority of the community to tip from destruction into economic and social viability. Viability means the likelihood of survival, and isn't survival what we're after?

Somewhere in the middle of the fellowship I hit a hard wall, culminating in my most embarrassing moment: crying uncontrollably on a coaching call with two of my fellowship advisors. It was hard to articulate then what I fully understand now—that hope, perseverance, human spirit, the odds, and a deep sense of protection toward wildfire survivors are barriers to turning this into a strictly academic exercise. I refuse to be compelled by data alone to say that our most vulnerable communities, hit hard by tragedy, will not survive because that's what the numbers say. Almost as heartbreaking as fire is, is predicting that some of the communities that burn are gone for good, and leaving it at that. Nothing saps my lifeblood faster than giving up on the human spirit.

I have to believe fire-impacted places will achieve a new community vision because of the people who love them. They may never return to exactly what they were before—perhaps not even to the conventional definition of viable—but people who love these communities, including government disaster workers, will breathe new life into them.

While touring the North Complex Fire burn scar three years after the fire, I stood behind a woman staring down into a ravine at the property she inherited from her grandmother. The land was charred as far as the eye could see, 100% tree mortality as the experts beside me said. She turned back to us and implored the group for help. The trees had already started halving themselves, breaking in the middle and casting off their tops. Trees fell into trees which fell into trees which fell into trees which slid down the steep slope in messy piles near her home at the bottom of the ravine.[5]

One by one the experts stepped up to talk about which trees to let fall, which to clear, and what to do with the dead wood. They talked about soil restoration, slope stabilization, and erosion control. They did not talk in months or even years, they talked in decades, in lifetimes. As it became clear that developing a restoration plan for her land would take far longer than a quick visit, she exchanged contact information and we left somberly in our caravan.

I can see why people who live far away from these circumstances shrug away the idea of recovery. I would if I still lived in the Bay Area. Why restore the dam when it breaks? Just tear it down. Why rebuild after wildfire? Just move away.

But as I stood there watching the landowner hold a sliver of hope, I could see how impossible it was for her to walk away. She loves the land that attaches her to her grandma, to her heritage, to her future. It's the same reason Melissa and her husband knew the remnants of their property were enough.

As much as I know money drives recovery, so does the unquantifiable, unbreakable, undeniable presence of love. It's why my parents won't leave the canyon. It's why I keep going back to Lake Pillsbury. I'm as certain of its power as I am of an adequate insurance pay-out's ability to fund reconstruction. And why focus on one and not the other?

---

[5] Butte County Fire Safe Council, North Complex Tour, 2024.

Love didn't come up in my interviews because it doesn't constitute a down payment, it can't master plan a development, it can't build a house on its own, but it might just be the most compelling variable in the mix when recovery involves choice. Love might explain why my friend who has moved three times since the Camp Fire recently told me she's rebuilding in Paradise. After years of suffering without love, she's moving home to find it.

Soft Spots

Many points I make in this book are observations of good work done around me during response and recovery. I am a student of recovery with each fire that impacts Butte County, learning from my wildfire elders and new staff who ask great questions. Hearing colleagues share personal and professional stories of response and recovery normalizes my own experience. In this way, counterintuitively, the EOC is a soothing place of shared history and common understanding.

Learning more about what breaks my heart with each successive fire helps me understand what kind of sadness I need to heal. Leaving behind shame for what I do and don't know, for what I can and can't do, makes room for sadness to move through and out.

Fire recovery is like putting together a delicate puzzle wearing a blindfold and heavy gloves. Thankfully, no one person is responsible for figuring it all out alone. That said, on a multi-agency recovery team there is tension and frustration, a little bit of fury, a fire lit inside of people who have to do it again and again at work. And, for the least fortunate in our community, at home, too.

I had hoped to wrap up the fellowship and this book in a neat bow. Alas, the communities I wrote about just a few pages ago, untouched by fire for decades, have now burned. I wrote about steeling myself for this happening again, and it did. I wondered if I have the courage to respond, and I do.

A Fire Captain recently explained that fire season used to line up with football season, fueled by autumn winds when tree leaves and grasses are dry. But that's not our experience in Butte County where conditions are ripe for wildfires to strike in early July, and LA where firestorms raged just days into 2025.

As we await the start of another fire year I wonder…how do we keep going?

To help myself through the emotional strain of fire, I now acknowledge my soft spots: the animals and the trees. I wear grief on my sleeve for everyone to see. Burying it in my heart allows it to sour and spoil, and grief that rots is toxic and immovable. My co-worker verbalized my particular pain on our Park Fire tour, and I allowed myself to double over in front of our Chief as if I'd been punched by fire. We have been punched, and this is where it hurts for me.

I don't know what adding new layers of wildfire means for me and my friendships, the wispy alliances that look more like smoke than string. We will see if I can rebuild those or if I should build new connections with people who are also steeped in fire and recovery. I'm surprisingly ok with the latter; the grief of my old life feels softer now that I've sifted through my own ashes and found the remains that matter most.

I have no shame about my sadness, or the fact that when I respond to work to help with these fires, they transform who I am. Trying to push my way to the front of this orchestra to wrest control of the conductor's wand feels like missing the point. The point is not to control this because that would be futile and maddening. The hard, gratifying work is accepting I have no control over these fires, and finding strength in my ability to show up for my community when called.

Storytelling

When we stood up the DROC for the Park Fire, I found myself driving to Paradise to deliver health and safety packets to a non-profit distributing them to survivors in Cohasset and Forest Ranch. I had the distinct feeling that Paradise is now the big sister to our newly recovering communities. I also felt disoriented that the County is once again walking survivors through the early steps of wildfire recovery.

A co-worker once worried aloud that we're like frogs in a pot of water slowly heating to a boil. What if we can't recognize the point at which we're suffering too much, and we never jump out? And does it matter if life and work carry on regardless, and we're compelled to help our communities through anything?

As I crested into Paradise on Skyway and looked at the Town rebuilding, I felt my mind and heart expanding to take on these new fires and our new reality. As insecure as I feel at work sometimes for how much I don't know, I also recognize in moments that I know more than most people outside of Butte County about wildfire and recovery through lived experience. Maybe that's enough.

In the coming months and years my goal is to find and strengthen connections with new friends who need special spaces for their stories, too. People who, by now, know what breaks their hearts during disaster and know what breaks mine. I need deep rest when I can get it and permission to write, write, write. Permission to process the pain by putting words to it, so it might ease the wildfire tumult in someone else's life.

My healing comes from turning storytelling into advocacy; helping draft letters the County sends to State and federal lobbyists and elected officials, and adding stories to the letters our NGOs write. The impact of wildfire on Butte County can be quantified, qualified, and should make a difference in the field of recovery, as policies and procedures are set, tested, and adapted. More than anything, it should be *felt* so it can be understood.

My fellowship mentor checked in with me during the Park Fire and I responded that my soul could sleep for weeks. While that was true as the fire raged out of control, I have limitless energy for digging in and articulating what does and does not work in disaster recovery. If there's a pathway for sharing, I believe I can keep showing up and doing the work.

Storytelling is the release valve for alienation, isolation, disorientation, rage, all the things we feel when fire arrives. Accepting this feels like my first act of personal resilience. This fellowship has been the doorway from personal suffering into healing, producing evidence of what can be felt but not seen in any of the research, revealing the one thread that ties together the incomprehensibility and impossibility of coming back after fire.

My definition of recovery is love.

GPSR Compliance
The European Union's (EU) General Product Safety Regulation (GPSR) is a set
of rules that requires consumer products to be safe and our obligations to
ensure this.

If you have any concerns about our products, you can contact us on

ProductSafety@springernature.com

In case Publisher is established outside the EU, the EU authorized
representative is:

Springer Nature Customer Service Center GmbH
Europaplatz 3
69115 Heidelberg, Germany